Make:
Getting
Started with
littleBits

**Ayah Bdeir and Matt
Richardson**

MAKER MEDIA
SAN FRANCISCO, CA

Make: Getting Started with littleBits

by Ayah Bdeir and Matt Richardson

Published by Maker Media, Inc., 1160 Battery Street East, Suite 125, San Francisco, California 94111.

Maker Media books may be purchased for educational, business, or sales promotional use. Online editions are also available for most titles (*http://my.safaribooksonline.com*). For more information, contact our distributor's corporate/institutional sales department: 800-998-9938 or *corporate@oreilly.com*.

Editor: Brian Jepson
Interior Designer: David Futato
Cover Designer: Joe Shouldice

April 2015:   First Edition

Revision History for the First Edition

2015-04-07:   First Release

See *http://oreilly.com/catalog/errata.csp?isbn=9781457186707* for release details.

978-1-4571-8670-7

[LSI]

Contents

Foreword by the Founder

You've already bought this book, so I'm not going to spend any time trying to convince you of the merits of the Maker Movement—my assumption is that you're sold. I assume you're sold on the idea that the gratification that one gets from making something—whether in robotics, 3D printing, or food—is very powerful. I assume you're sold on the promise of social, economic, and educational change that can come from spreading the ethos of STEM/STEAM and "learning by making." I assume that you believe that promoting a society where people make, remix and share online or in social settings—as opposed to a silo culture—is a good thing.

But I'm writing today to talk about an even bigger idea: the idea that we don't just need to be makers, we need to be inventors. I believe that inventors are an evolved breed of makers. Inventors are sometimes problem-solvers that are inventing a solution to a particular problem, and other times they are creative thinkers that are inventing a future that has never been imagined. To elevate making to inventing, we need to equip ourselves with a new language to understand the world around us, and a platform to reinvent it.

We spend more than 11 hours with electronic devices every single day, but most of us don't know how they work, or how to make our own. When I first started working on littleBits in 2008, this number was 7.5. Technology has moved from being an integral part of our lives, to helping define who we are. It's the cars we drive, the phones we own, the alarm systems that keep us safe, the iPads that two year old kids tap and swipe. Yet engineering is mysticized, electronic objects are black-boxed, and if we are honest with ourselves, we have ceased to understand the technological world we live in. And in the meantime, the world is moving at a very fast pace, from the Internet of Things to Artificial Intelligence, each with their own promises and challenges. How can we solve the challenges we face today if we don't understand the world we live in? I believe that to solve 21st century challenges—economic, environmental, medical—we don't need more, we need smarter. Tomorrow's sense of pride will come from inventing the future.

As Matt and I were writing this book, I reviewed some of the very early images, sketches, and writings from the first days of littleBits as a project. I am incredibly humbled and surpassingly excited about what lies ahead. I hope you enjoy this book and all the love and care that has gone into it, and into littleBits as a product and a company. I would like to deeply thank our team of bitsters who work tirelessly to make this vision come to life. They are some of the most talented and dedicated people I have ever had a chance to collaborate with, and they keep it EPIC. But most of all, I would like to thank the littleBits community, an incredibly diverse community of artists, designers, kids, engineers, hackers, educators, librarians, from all ages, all languages, and interests. You never cease to amaze me with all the stunning, shocking, and delightful inventions you come up with every day.

Now enough of me talking, go ahead and start inventing!

— Ayah Bdeir, Founder and CEO, littleBits

Preface

I started working on littleBits in early 2008. I had graduated from the MIT Media Lab and had a prior background in Electrical Engineering. I grew up in Beirut, Lebanon, and to be honest, I never wanted to be an engineer. I am one of four girls and my parents tell me I was always a tinkerer, a maker and very often, a breaker. When it was time for me to decide what I wanted to do in university, my parents and teachers said I owed it to myself to be an engineer because I was good at Math and Science. But I had always found that engineering was dry and not creative. It wasn't until I went to the Media lab that I discovered the power of engineering when combined with creativity. I started to create my own artwork using electronics: wearable electronic fashion, interactive installations, lighting art. A little while after, I realized I was more interested in the tool than the outcome of what I was creating. I had been working with the design agency Smart Design with a colleague Jeff Hoefs, and we wrote a paper called "Electronics as Material." Together we designed some of the earliest prototypes of littleBits and this was the beginning of a long research to try to put the power of electronics in the hands of everyone.

My two biggest inspirations were Lego and Object-Oriented programming, two of the most successful modular systems of our time.

Modular Electronics

To understand complex ideas, I believe in the power of modular systems. Modularity allows us to understand complex notions that we may have previously found intimidating, by allowing us to break them down, and build an even more complex idea back up, one building block at a time.

The first inspiration was a big one. In 1947, Lego had managed to take the cement brick, the most important construction unit in the world, and make it an imagination tool, accessible to everyday people. With Lego you didn't have to be an expert to make a complex structure, you learnt intuitively, and could build more and more sophisticated structures one brick at a time. In a few short years, Lego bricks took place in every household. It is estimated that over 400 billion bricks have been produced, or 70 bricks for every person on the planet. We didn't have to be engineers to make walls, houses, buildings, bridges. Lego had taken the building block of our time and made it into the building block of our imagination. Suddenly we gained an understanding of the world around us. Structures that we saw on the streets of city centers that previously appeared huge and complicated were not so unattainable, not so intimidating anymore: you could clearly imagine yourself building them up, one brick at a time.

Figure P-1. *One brick at a time (graja/Shutterstock)*

Figure P-2. *Croatian National Theater building made of Lego blocks (Gordana Sermek/Shutterstock)*

The second inspiration was Object-Oriented Programming. Software used to be linear, obscure, and thus only reserved for experts. Then Object-Oriented Programming came along. It introduced the concept of modular blocks, allowing people to reuse pieces of code written by them and other people, and build more and more sophisticated code, one brick at a time. Now anyone with two weeks and a computer can learn to make the most successful game in the world.

But in hardware, this is still not possible. The hardware industry is a very top-down industry where prototyping times are long, expertise is required, and the field in large part still belongs to engineers. So how do we put the power of electronics in hands of everyone? We make electronics modular.

I built the first littleBits prototypes using cardboard, devising a technique using copper tape from Home Depot. It was the best way to touch, and feel the modules. To imagine how a person who had never touched electronics before would interact with them, how they would be inspired by them.

Figure P-3. *One of the first prototypes of littleBits from 2008, using copper tape from Home Depot*

Figure P-4. *Another prototype. Note the ring magnets held together with metal pins*

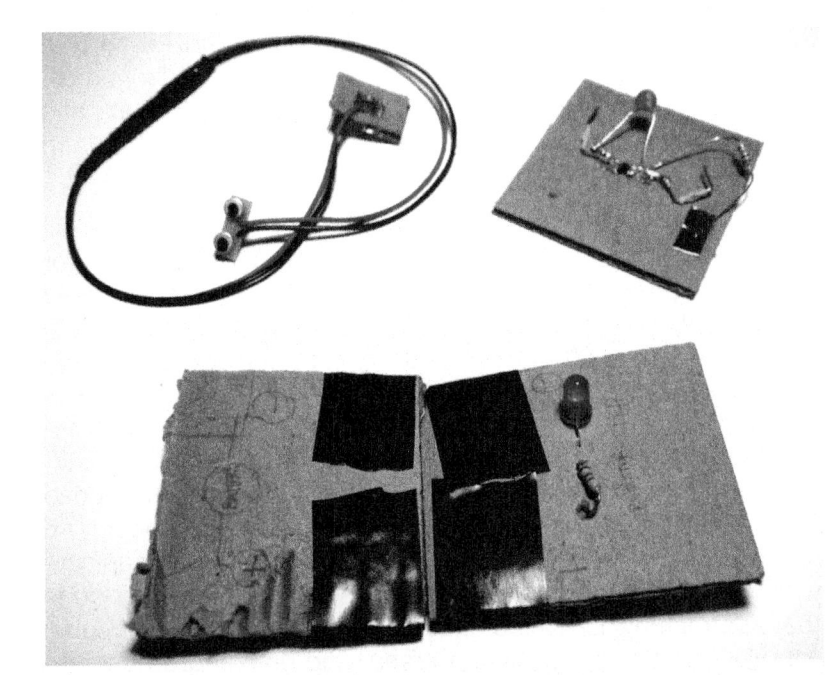

Figure P-5. *Various prototypes ready to be snapped together!*

Every single aspect of littleBits was up for design, nothing was taken for granted. Over 3.5 years there were hundreds if not thousands of experiments and decisions that led to littleBits as it is today. First and foremost of course, electronics design, so that the system of circuits could be genuinely modular: any Bit has to work with any other Bit in the system, and the library should be infinitely extendable. It was also extremely important to figure out the right level of abstraction for each module. little-Bits are not component-level modules, they are block-diagram level modules. Nailing how high-level the block diagram had to be, in order to make sure it is understandable, but also how low-level so that it can be versatile, was crucial.

But beyond electronics design: interface design, mechanical design, cost, branding, aesthetics, naming convention, color code—every single aspect was a process, and a decision. For example: I searched for connectors that were easy to attach and detach for weeks, it was very important to take any fear or uncertainty out of making electronics. The connectors needed to be small, iterative, but most importantly, polarized so that you couldn't make any mistakes and do something dangerous. After hundreds of connectors, I settled on something that I had never yet seen put into electronic circuits before: magnets. And they had the added benefit of making anything feel magical.

The size of the circuits needed to be designed in multiples so that larger circuits could work in any configuration and allow for 2 and 3D rotation, even if that meant sometimes modules would be a little bigger or smaller than they ideally wanted to be. I wanted the circuits to appear inviting, not intimidating as green and black PCBs often are, so I tried different circuit board colors to look crisp and clean—white was the way to do that. The circuits needed to feel human and gender-neutral, hence the handwritten font to denote the name of the module, and they needed to be a building block for creativity, not a finished product, so the circuitry is exposed. There are lots of rules to communicate when it comes to rules of electronics, so we made a color-code that abstracted the rules of electricity into a manageable code: you always need a blue and a green, and pink and orange are optional, in between. I needed to make sure you could understand immediately what each module was and how to interact with it, so the user interface puts any interaction point at the top, and all other circuitry at the bottom, even though it's not always the most sound way to design a circuit.

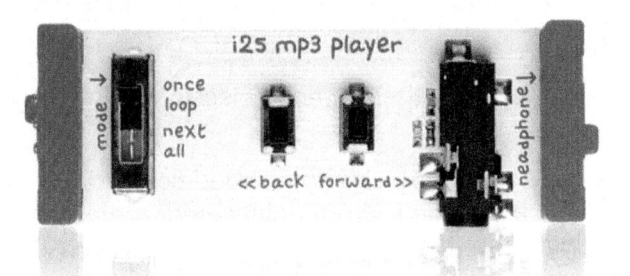

Figure P-6. *After thousands of experiments and decisions, the littleBits module and design language is born*

I can go through every single aspect of the modules and the system, but it would be a whole other book. Trust me when I say that no aspect of the engineering, design, mechanics, manufacturing or interaction of littleBits is arbitrary; every single aspect is deliberate. But the good news is, all it takes is to see someone pick them up, whatever their age, gender, background or country of origin, snap the first Bits together with no instructions and see a light come on. Right then and there, you see their face lights up, and you know you've created something truly universal.

Fast forward 3.5 years and 27 prototypes later, the littleBits library was born. A first-of-its-kind modular electronics platform for learning and inventing.

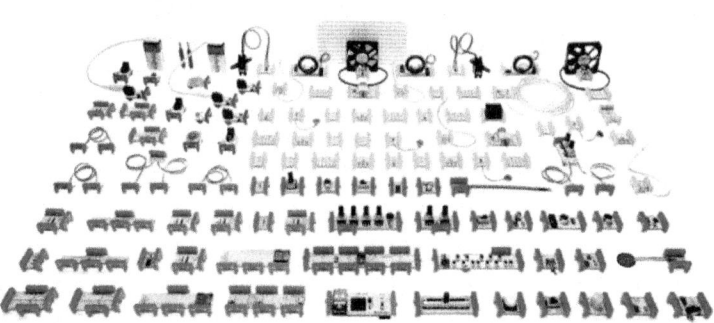

Every Interaction is a Ready-to-Use Brick

Our mission is to put the power of electronics in the hands of everyone.

littleBits is a library of electronic modules for learning and inventing. Each module is a pre-engineered, pre-assembled circuit ranging from the very simple (lights, sounds, sensors, motors) to the very complex (wireless radio frequency, programmability, cloud connectivity). Modules snap with magnets so you can't put them the wrong way, with no soldering, wiring or programming, unless you want to. The modules are color-coded: blue is power, pink is input, green is output, and orange is wire. All you need is a blue and a green; pink and orange are optional in between. There are billions of combinations of circuits possible, and the library is infinitely extendable. The littleBits library is open source, and fosters a community of contributors that can redesign, share online, and learn from each other's ingenuity.

Through looking at the world through the lens of the littleBits modular platform, you can break down any complex electronic device and understand how it works. Now you have a language to understand the world around you, from simple dimmer lights, to automatic elevator doors to internet-connected thermostats.

Figure P-7. *A lightbulb deconstructed: power, bright led*

Figure P-8. *A nightlight: power, slide switch, light sensor, bright led*

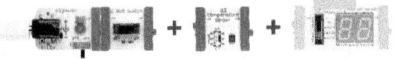

Figure P-9. *A digital thermostat: power, slide switch, temperature sensor, number*

Figure P-10. *A DIY Nest: power, fork, button, temperature sensor, dimmer, Arduino, latch, bright led, number, servo, cloud*

DIY electronic kits are a dime a dozen, but what we are most interested in is a language to allow you understand the world we live in, and to reinvent it. That way we can encourage people to look around them, question the devices and phenomena they take for granted, and get inspired for their next invention.

Today, littleBits (*https://youtu.be/YUUsJSDa7PE*) is a library of over 70 Bits modules and hundreds of billions of possible combinations. We've designed the system carefully to make it gender-neutral, age-agnostic and independent of discipline or technical experience. With these Bits modules you can make things that would otherwise require programming, soldering, and complex microcontrollers. You can make circuits with timing functions and logic that rival the most complex robotics tools. Over the past 3 years we reached outside the choir and enabled people who never thought of themselves as "makers" to jump in and create their own inventions with electronics. No matter if you were a 30-year-old designer from New York (*http://littlebits.cc/designer-spotlight-ron-rosenman*), an eight-year-old boy from Singapore (*http://littlebits.cc/tag/chow-yu-hin*) or an educator from NASA (*http://littlebits.cc/maker-spotlight-ginger-butcher*), we set out to enable you to learn and invent with electronics within seconds: from a connected door-bell (*http://youtu.be/hJsrma74Lns*) to a fully responsive robotic installation (*http://youtu.be/ij0toovW9HI*).

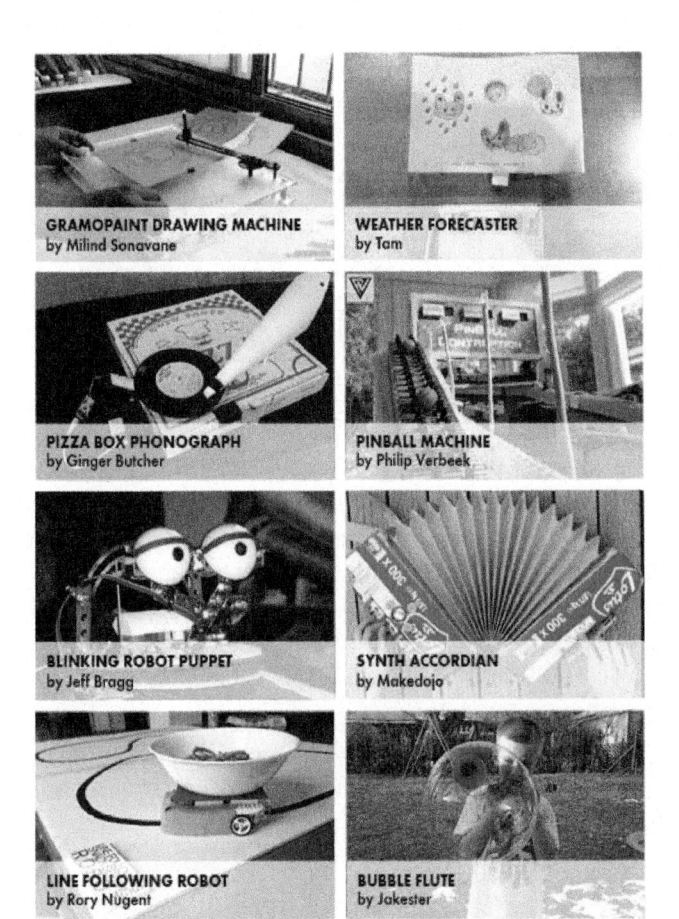

GRAMOPAINT DRAWING MACHINE
by Milind Sonavane

WEATHER FORECASTER
by Tam

PIZZA BOX PHONOGRAPH
by Ginger Butcher

PINBALL MACHINE
by Philip Verbeek

BLINKING ROBOT PUPPET
by Jeff Bragg

SYNTH ACCORDIAN
by Makedojo

LINE FOLLOWING ROBOT
by Rory Nugent

BUBBLE FLUTE
by Jakester

Today littleBits are used in over 3,000 schools, afterschool programs and libraries in over 70 countries, and have enabled educators teaching across all grade levels and topics, from grammar (*http://clmoocmb.educatorinnovator.org/2014/assign ments/learning-grammar-with-little-bits/*), to music (*http://little bits.cc/browse-lessons/vocal-synthesizer-workshop-modeling-the-human-vocal-tract*) to 21st century product design (*http:// littlebits.cc/browse-lessons/how-to-run-a-littlebits-workshop*).

Over the past years we have had the pleasure of collaborating with some of the most prestigious design and technology leaders in the world: MoMA (*http://youtu.be/K4fjKQPZ62Y*), KORG (*http://youtu.be/vDxz8bpqOMU*) and NASA (*http://youtu.be/WrywrtSnSog*) to name a few and we are stopping at nothing. We are extending the invention platform far and wide. First, technologically: with the bitLab (*http://littlebits.cc/bitlab*), anyone can design their own modules, get the community to vote, and we will manufacture and share the proceeds.

And then geographically: this year we launched a global chapter program to enable partner makerspaces, schools and design studios to help spread our mission and run programs and workshops all over the world.

As you are going through this book, make sure to frequently visit littleBits.cc (*http://littlebits.cc/*) to meet other members of the community, get inspired by their inventions, and get access to the latest lessons. And of course, follow us on instagram (*http://instagram.com/littlebits*) and twitter (*https://twitter.com/littlebits*) and show off your own inventions!

— Ayah Bdeir, Founder and CEO, littleBits

Conventions Used in This Book

The following typographical conventions are used in this book:

Italic

Indicates new terms, URLs, email addresses, filenames, and file extensions.

`Constant width`

Used for program listings, as well as within paragraphs to refer to program elements such as variable or function names, databases, data types, environment variables, statements, and keywords.

`Constant width bold`

Shows commands or other text that should be typed literally by the user.

`Constant width italic`

Shows text that should be replaced with user-supplied values or by values determined by context.

 This element signifies a tip, suggestion, or general note.

This element indicates a warning or caution.

Using Code Examples

This book is here to help you get your job done. In general, you may use the code in this book in your programs and documentation. You do not need to contact us for permission unless you're reproducing a significant portion of the code. For example, writing a program that uses several chunks of code from this book does not require permission. Selling or distributing a CD-ROM of examples from Make: books does require permission. Answering a question by citing this book and quoting example code does not require permission. Incorporating a significant amount of example code from this book into your product's documentation does require permission.

We appreciate, but do not require, attribution. An attribution usually includes the title, author, publisher, and ISBN. For example: "*Getting Started With littleBits* by Ayah Bdeir and Matt Richardson (Maker Media). Copyright 2015, 978-1-4571-8670-7."

If you feel your use of code examples falls outside fair use or the permission given here, feel free to contact us at *bookpermissions@makermedia.com*.

Safari® Books Online

Safari Books Online is an on-demand digital library that delivers expert content in both book and video form from the world's leading authors in technology and business.

Technology professionals, software developers, web designers, and business and creative professionals use Safari Books Online as their primary resource for research, problem solving, learning, and certification training.

Safari Books Online offers a range of product mixes and pricing programs for organizations, government agencies, and individuals. Subscribers have access to thousands of books, training videos, and prepublication manuscripts in one fully searchable database from publishers like Maker Media, O'Reilly Media, Prentice Hall Professional, Addison-Wesley Professional, Microsoft Press, Sams, Que, Peachpit Press, Focal Press, Cisco Press, John Wiley & Sons, Syngress, Morgan Kaufmann, IBM Redbooks, Packt, Adobe Press, FT Press, Apress, Manning, New Riders, McGraw-Hill, Jones & Bartlett, Course Technology, and dozens more. For more information about Safari Books Online, please visit us online.

How to Contact Us

Please address comments and questions concerning this book to the publisher:

Make:
1160 Battery Street East, Suite 125
San Francisco, CA 94111
877-306-6253 (in the United States or Canada)
707-639-1355 (international or local)

Make: unites, inspires, informs, and entertains a growing community of resourceful people who undertake amazing projects in their backyards, basements, and garages. Make: celebrates your right to tweak, hack, and bend any technology to your will. The Make: audience continues to be a growing culture and community that believes in bettering ourselves, our environment, our educational system—our entire world. This is much more than an audience, it's a worldwide movement that Make: is leading—we call it the Maker Movement.

For more information about Make:, visit us online:

Make: magazine: *http://makezine.com/magazine/*
Maker Faire: *http://makerfaire.com*
Makezine.com: *http://makezine.com*
Maker Shed: *http://makershed.com/*

We have a web page for this book, where we list errata, examples, and any additional information. You can access this page at: *http://shop.oreilly.com/product/0636920032038.do*.

To comment or ask technical questions about this book, send email to: bookquestions@oreilly.com

1/littleBits Basics: Inputs and Outputs

There's no better way to learn how to use littleBits than to jump right in and try them out. In this chapter, you'll cover all the basics: how to power your Bits, how they connect—and you'll look at a few of the different inputs and outputs that you can use in your projects.

The Bits

While there are over 60 different modules (or *Bits*) in the little-Bits library to choose from, every module falls into one of four different categories. Each category has a particular color to make the modules easy to find and identify:

- Power (blue)
- Output (green)
- Input (pink)
- Wire (orange)

Every Bit works with every other Bit in the library and it can keep growing to infinity. You can even create your own Bits! But more on that in Chapter 6.

The Bits connect to each other magnetically with their *bitSnap* connector. This unique feature of the Bits helps you easily make the physical and electrical connections so that you can focus on creating your project. That means that there's no need to worry about soldering or making sure you're connecting the right wires. You'll learn more about them in "Under the Hood: bitSnap Connectors" on page 13.

Figure 1-1. *The bitSnap connector makes it easy to connect modules together.*

This chapter focuses on your first steps with each of the types of modules and a few littleBits accessories.

 Most of the examples in this chapter will require the modules in the Base Kit (*http://littlebits.cc/kits/base-kit*), but for maximum fun, consider the Premium Kit (*http://littlebits.cc/kits/premium-kit*) or Deluxe Kit (*http://littlebits.cc/kits/deluxe-kit*), which include more Bits and accessories. Some of the projects will require other Bits as well.

If you don't have a particular Bit, you can often make substitutions. For example, the Base Kit includes the dimmer Bit (i6) that's used later in this chapter. The Premium Kit and Deluxe Kit include a fancier dimmer, the slide dimmer Bit (i5) that works just as well.

The (Only) Two Rules of littleBits

1. The magnets are always right.
2. You always start with a blue and a green; pink and orange are optional (in between).

Power (Blue)

Of course, electronics need electricity, so every project you make with littleBits is going to start with a blue power module. Most commonly, you'll encounter the power module (module p1) pictured in Figure 1-2. It takes in 9 to 12 volts from a battery or power supply and converts it to the 5 volts used in littleBits circuits. It has a built-in switch that you can use to turn your project on and off. There's also an on-board LED to indicate when power is being supplied to your project.

> All of the modules in the littleBits library work with 5 volts of electricity. No matter what type of power module you use, it will ensure that your circuit is supplied with 5 volts.

Figure 1-2. *Power module P1 lets you use a 9 to 12 volt DC input to power your littleBits project.*

There's a cable that will connect a 9 volt battery to the power module (as shown in Figure 1-3), but if you want to power your project from a wall outlet, you could also use a DC "wall wart" adapter with a 2.1mm inner diameter barrel jack adapter (center pole positive), which is the same size of connector used for the Arduino Uno. Just be sure that the output of the adapter is between 9 and 12 volts direct current (DC). The output voltage will be listed on the adapter, as shown in Figure 1-4.

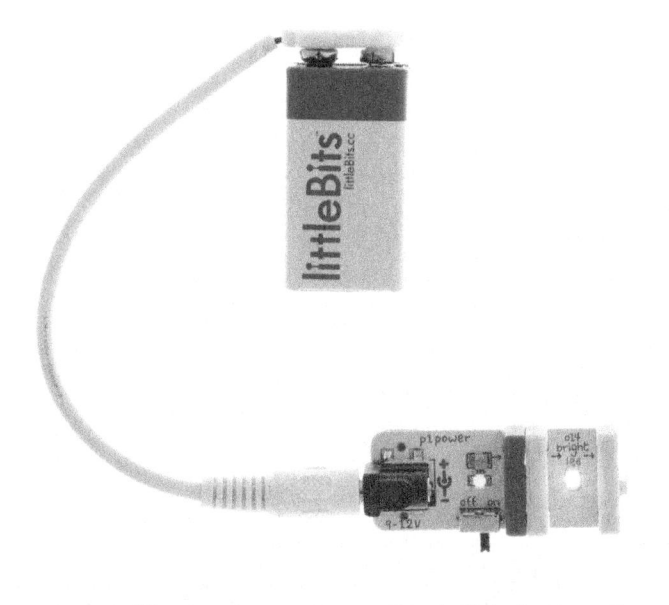

Figure 1-3. *The basic power module (p1) is frequently used for connecting a 9 volt battery to a project.*

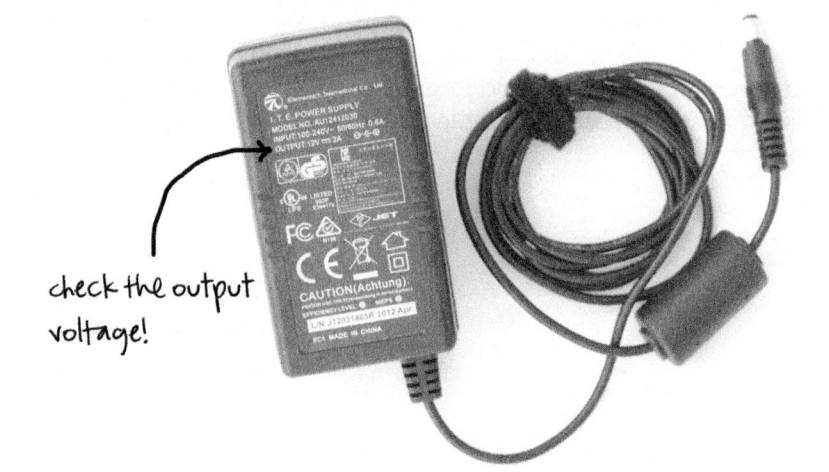

check the output voltage!

Figure 1-4. *This wall wart power supply has a 12 volt direct current output.*

Under the Hood: Voltage Regulator

One of the core parts on the power module is the *voltage regulator*. A common component in electronics projects, the voltage regulator's job is to take a range of *input voltages* and supply a specific *output voltage*. In the case of the voltage regulator on the power module, it can take 9 to 12 volts as an input and it provides 5 volts as output. The voltage regulator, along with a few other components, ensure that all the Bits in your project receive a steady 5 volts.

Outside of the power module included in the Base Kit, you have a few other options for powering your project. Most of the time, you can use whichever power module you'd like. We'll note a few exceptions later in the book.

For example, if you need something even more portable than a 9 volt battery, you could also use the coin battery power module (p2), which is a small rechargeable power source as shown in Figure 1-5. It will power most littleBits, but any mechanical Bits will deplete the battery quickly. Like the p1 module, it has a switch to turn it on and off. When it's out of juice, plug it in with a micro USB cable to charge it up until the light goes from yellow to green and you'll be good to go again. The coin battery's small size makes it great for portable and wearable projects.

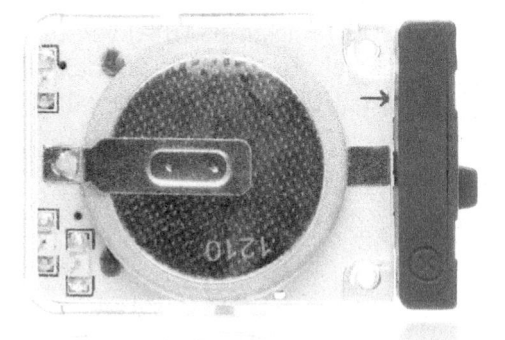

Figure 1-5. *The rechargable coin battery module (p2) is perfect for wearables.*

Under the Hood: Boost Converter

The coin battery power module uses a 3.3 volt battery inside. Because all littleBits work with 5 volts power, the circuitry on the module includes a *boost converter*, which raises the voltage to 5 volts to be compatible with the other littleBits modules.

You can also power your littleBits directly from any USB power source with the USB power module (p3), shown in Figure 1-6. If you have any spare device chargers with a USB output, use a micro USB cable to connect them to this module and your project is powered. Or connect a USB battery pack (the kind that charge up your phone in a pinch) to get power on-the-go. If you use your computer's USB ports to power your project, just keep in mind that it's only using the power from your computer; there's no data connection being made with this module.

 Not all USB power supplies are created equal. If you're in doubt, we recommended that you use the official littleBits USB power supply, which supplies a generous 2000 milliamps (mA) versus the 500 mA that you're likely to get from your computer's USB ports or the 1000 mA common with many device chargers.

Figure 1-6. *Power your Bits via USB with the USB power module (p3).*

Some Bits, such as the cloudBit covered in Chapter 4 *require* you to use the USB power module. Because the cloudBit requires significant power, you should be sure to use the littleBits wall adapter supplied with the USB power module.

As you can see, there are many options for powering your project. As long as you have one of these power modules and a power source, you're ready to go with littleBits.

Now let's take a look at what you can do with that power.

Output (Green)

The green output are how your project will make stuff happen. It could be making light, motion, or sound. They let you see, feel, and hear your project.

For instance, the bright LED module (o14) use the electrical power from a power module to make light. Try it out now (Figure 1-7):

1. Snap the LED module to a power module.
2. Ensure that the power module's switch is turned on.

If you see the LED light up, congratulations! You just made your first circuit with littleBits! Pretty easy, right?

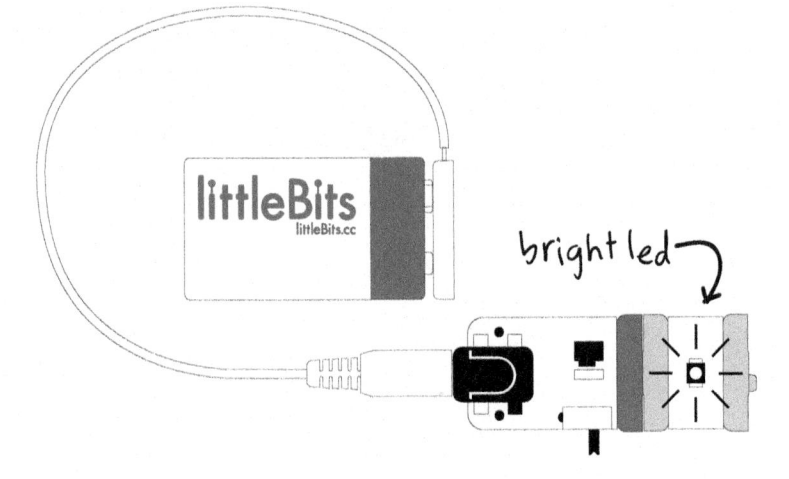

Figure 1-7. *Connecting a bright LED Bit (o14) to power (p1).*

You may have noticed that there's only one direction that the output LED module will connect to the power module. There are a few ways to know that you've got the connection right:

- The magnets between the bitSnap connectors will attract each other if you're connecting them the right way. If you feel a repelling force when you try to snap them together, you'll know you need to flip the LED module.
- The arrows printed on the Bit will point in the direction of the flow of power from the power module
- The top of the module has the part number and name of the Bit. The bottom of the module has the Open Source Hardware logo and the littleBits logo (three circled X's).
- The circled X symbol on the colored bitSnap connectors will line up with each other.

Each output module uses the electrical power it needs from a power module and also acts as a connector between power and other modules. Subsequent modules that you connect to the output module will also receive power.

1. Now try adding another green output module to the right side of the LED module to make a chain of outputs.

Other output modules (Figure 1-8) include motors, buzzers, speakers, and colorful LEDs, to name a few. This section doesn't cover every possible output module (because there are a lot!) but look for other output modules to make their entrance later on in the book.

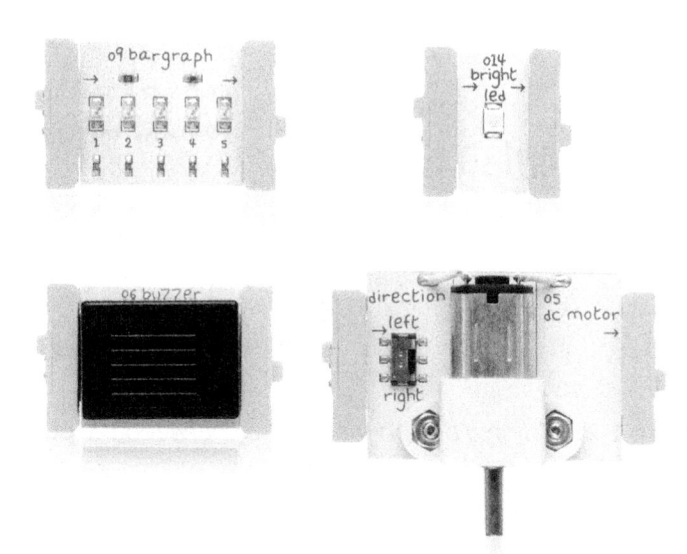

Figure 1-8. *These are the output modules that you'll find in the Base Kit: bargraph (top left), bright LED (top right), buzzer (bottom left), and DC motor (bottom right). There are many different output modules in the full littleBits library!*

Input (Pink)

The pink modules act as the inputs to a project. They allow you to interact with it and let your project interpret its surroundings. For example, a user can press a button to start a sequence of events, or your project may monitor the ambient light level in a room and react when it gets dark, just like Konstantin Bauman did with his Night Airplane project, which is featured at the end of this chapter (See "Project: Night Airplane" on page 28.).

Try out your first input module now (Figure 1-9)!

1. Connect the button module (i3) to a power module.
2. Connect a bright LED module (o14) to the button module.
3. Press the button.

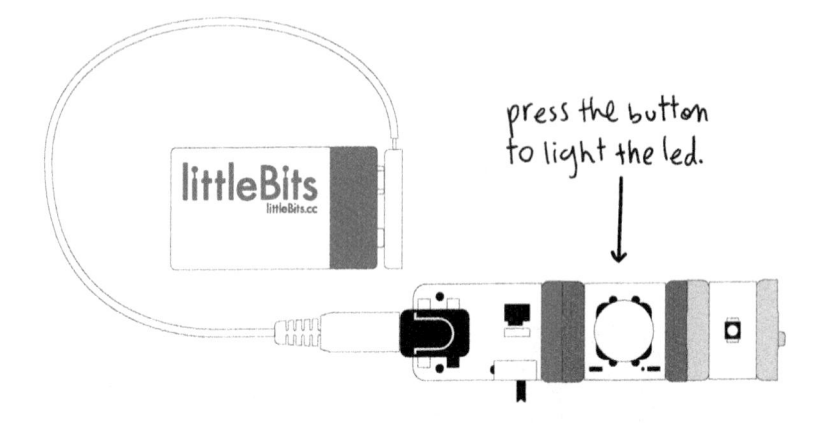

Figure 1-9. *Adding a button module (i3) between the power and bright LED Bits.*

If you got it right, when you press the button, the LED will turn on. If not, make sure that your power module is connected to a battery or other power source and the switch on the power module is turned on.

When you press the button, it sends a 5 volt signal from the power module to the LED module. When you let the button go, it breaks that connection, the signal goes to 0 volts, and the LED turns off.

Under the Hood: bitSnap Connectors

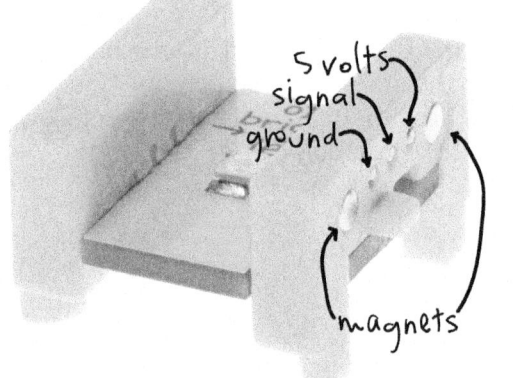

Let's take a closer look at how littleBits work.

If you look at the end of the littleBits bitSnap connectors, you'll see that there are five metal pads. The two outer pads are actually the magnets which hold the Bits together. The three inside pads are electrical terminals. The middle terminal carries the signal, which is how the Bits communicate. The signal can be anywhere from 0 to 5 volts. 0 volts is an OFF signal, whereas 5 volts is an ON signal.

The terminals on either side of the signal terminal provide power to the Bits. Among the two power terminals, the one that's closer to the circled x symbol on the top of the bitSnap connector is ground and the other power terminal is connected to 5 volts.

The signal terminal's voltage affects how the output Bits behave. For example, the more voltage there is on the signal line going into the LED Bit, the brighter the LED will be. The input Bits let you change the signal line's voltage and affect the Bits that come after it.

Now try this (see Figure 1-10):

1. Connect an LED to a power module.
2. Connect a button to the LED.
3. Connect a second LED to the button.

As you'll see, the LED module before the button stays illuminated no matter the state of the switch. An input module will only affect the modules that come *after* it when following the signal flow from the power module.

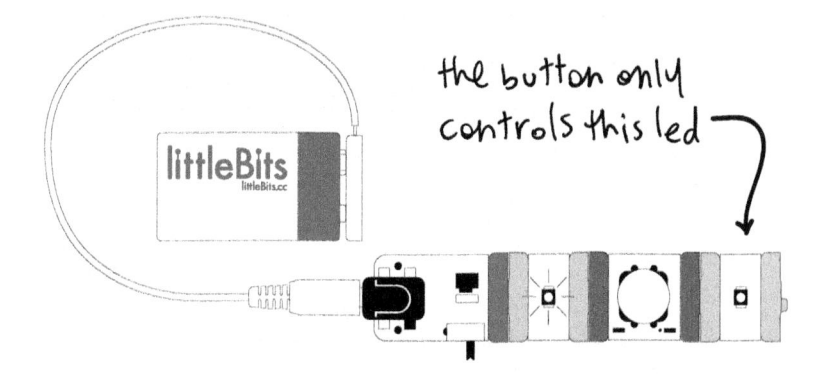

Figure 1-10. *The first LED always stays on*

The button represents a digital input and it can only be either on or off, never anything in between. In the littleBits library, some of the input modules are digital such as the switch (i2), motion trigger (i18), roller switch (i19), and pulse (i16).

There are also many modules that work as analog inputs and allow for a range of values between 0 and 5 volts. If you swap the button for the dimmer (i6), you can see the effect of changing the voltage. To create the circuit, shown in Figure 1-11, from scratch:

1. Connect a dimmer to a power module.
2. Connect the LED module to the dimmer module.
3. Try adjusting the dimmer and watch its effect on the LED.

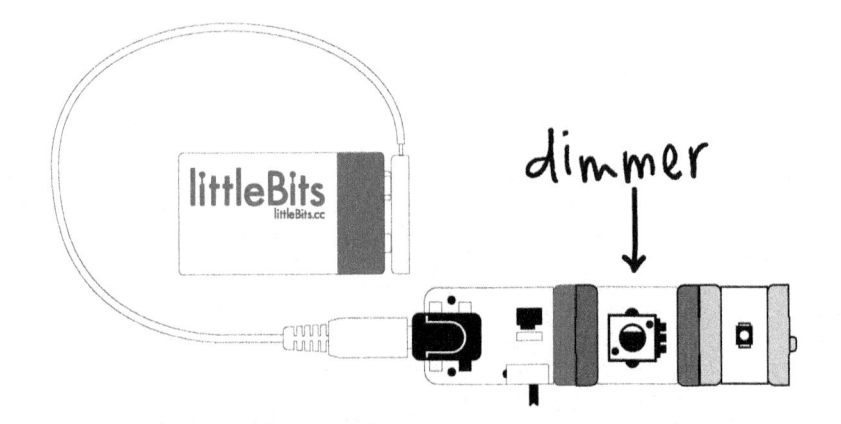

Figure 1-11. *Trying an analog input, the dimmer (i6).*

Just as with the button, the analog inputs will only have an effect on the Bits that are placed *after* them (follow the signal flow from the power module).

Figure 1-12. *These are the input modules that you'll find in the Base Kit: button (i3), light sensor (i13), and dimmer (i6). There are over 25 different input modules in the full littleBits library!*

That should help make the difference between the dimmer (analog) and the button (digital) clear. Now is a good time to try out some of the other output modules you have to see how they respond to an analog input.

1. Building from the previous circuit, connect a bargraph module to the LED module.

2. Adjust the dimmer and watch its effect on the bargraph output module.

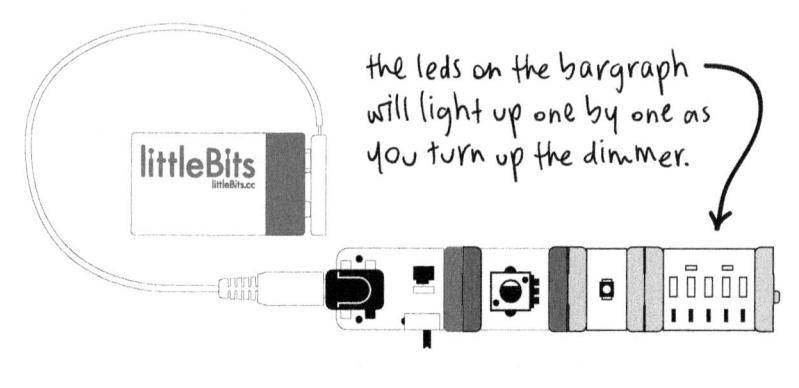

the leds on the bargraph will light up one by one as you turn up the dimmer.

Figure 1-13. *Trying the bargraph Bit, which behaves differently than the LED in response to analog input.*

As you turn the dial, watch its effect on the bargraph module. Try the other analog input Bit from the Base Kit, the light sensor (i13) to see its effect. You may need to use the included screwdriver to adjust the sensitivity of the light sensor Bit.

An analog signal is capable of a range of voltages whereas the digital signal is only either totally OFF (0 volts) or totally ON (5 volts). The graphs in Figure 1-14 show the differences in the signals over time.

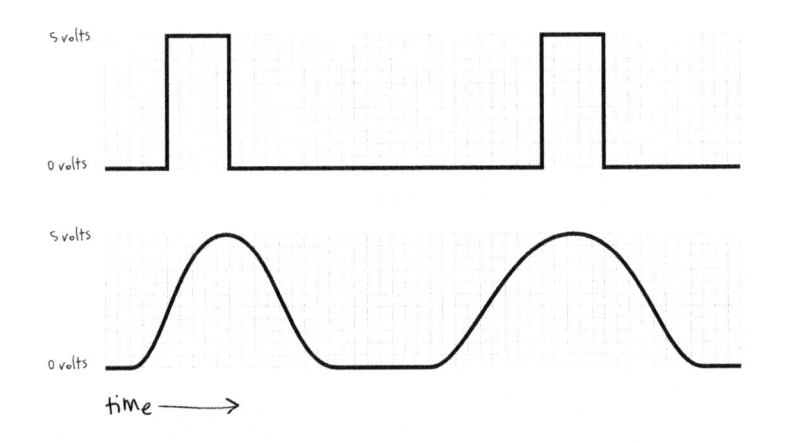

Figure 1-14. *The top graph shows a digital signal, one that's either OFF or ON. An analog signal, like the one on the bottom graph, can have a range of values.*

Wire (Orange)

The orange wire modules expand the ways that you can connect Bits together. They add flexibility to how you arrange the Bits and how they interact with each other. Some wire Bits enable you to add digital logic (covered in Chapter 2), Internet connectivity (covered in Chapter 4), and programmability (covered in Chapter 5) to your projects.

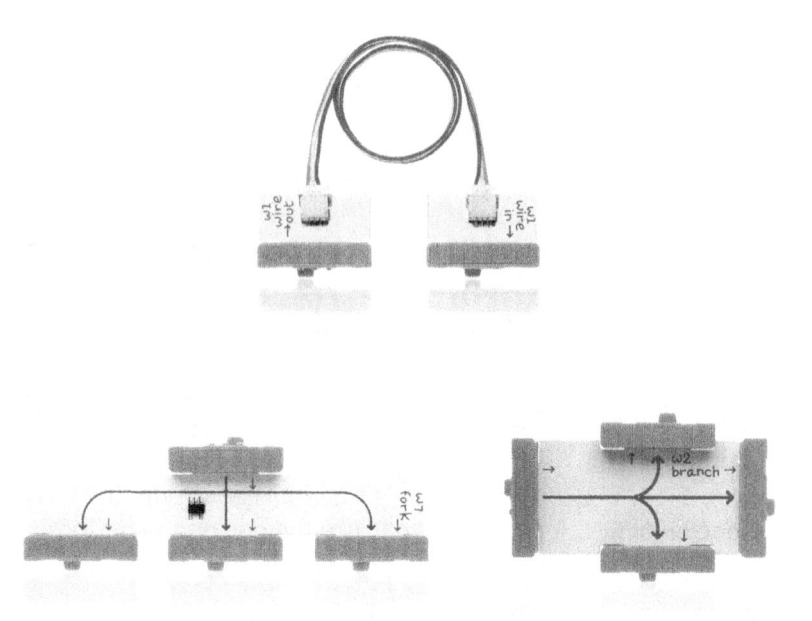

Figure 1-15. *The wire, fork, and branch modules help you connect different Bits together. These are the three most basic wire modules.*

There are two wire modules in the Base Kit. Both are called wire Bits (w1), which allow you to physically separate your Bits, as shown in shown in Figure 1-16. They will come in handy when you need to adjust the arrangement of the inputs and outputs. For instance, if you want to be sure a button is placed in just the right spot on your project, simply place a wire bit before and after the button and you'll have the flexibility to place the button anywhere.

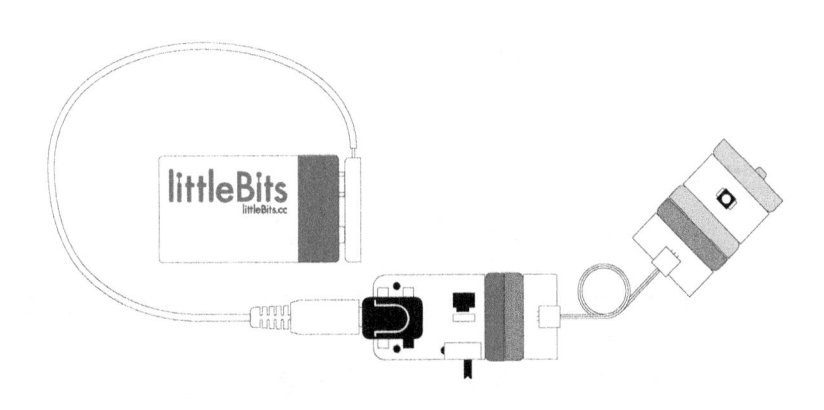

Figure 1-16. *The wire Bit (w1) allows you to physically separate your Bits.*

Outside of the Base Kit, there's a pair of simple but handy wire Bits called the branch (w2) and fork (w7) Bits, shown in Figure 1-17 and Figure 1-18. Both work the same way: they let you connect the output of a single Bit to as many as three other Bits. You could connect one of these Bits to a power module and then the output of that module can go to three different input and output Bits.

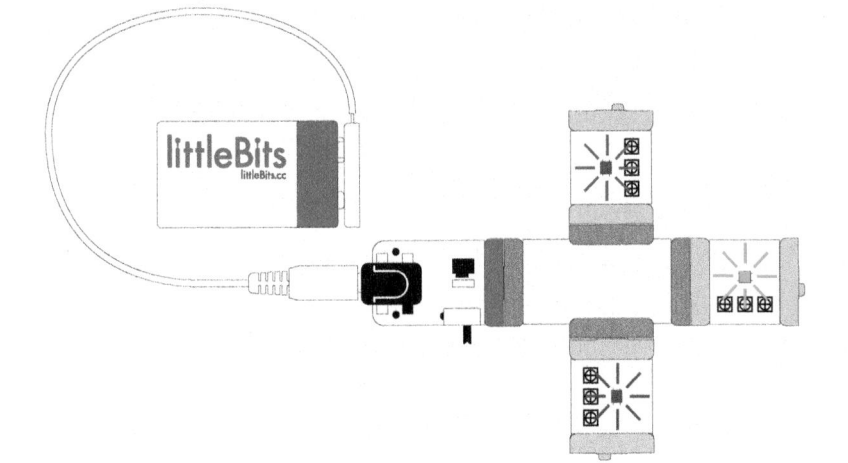

Figure 1-17. *The branch Bit (w2) lets you connect the output of a single bit to as many as three others*

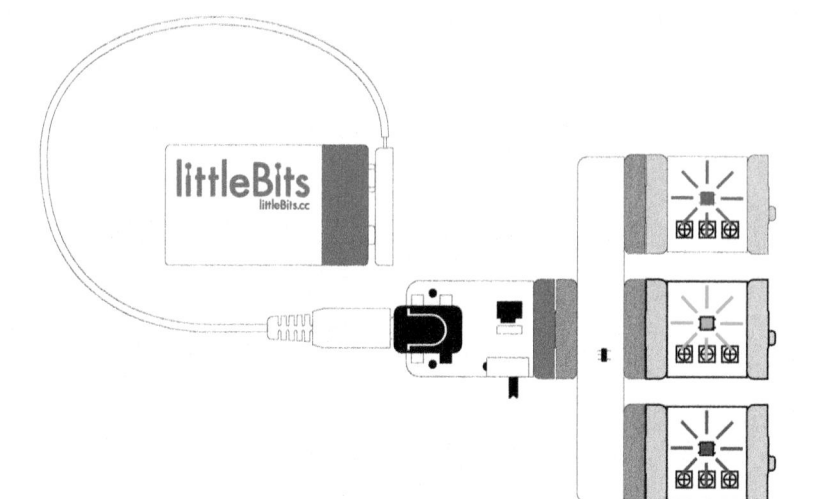

Figure 1-18. *The fork Bit (w7) works the same way as the branch bit, but is in a different physical configuration.*

If you want a button to control a fan but a dimmer to control an LED, you'll need to use the branch or fork Bits so that each input and output pair can have independent control. If you have a branch or fork Bit, try out out the circuit shown in Figure 1-19:

1. Connect the branch module to a power module.
2. Connect a button to one of the outputs of the branch.
3. Connect any output module like the fan (o13) to the button.
4. Connect a dimmer to another output of the branch.
5. Connect an LED to the output of the dimmer.

In this setup, the button can activate the fan without affecting the LED. The dimmer can fade the LED without affecting the fan. Without the branch or fork Bits, you wouldn't be able to do this while using a single power supply.

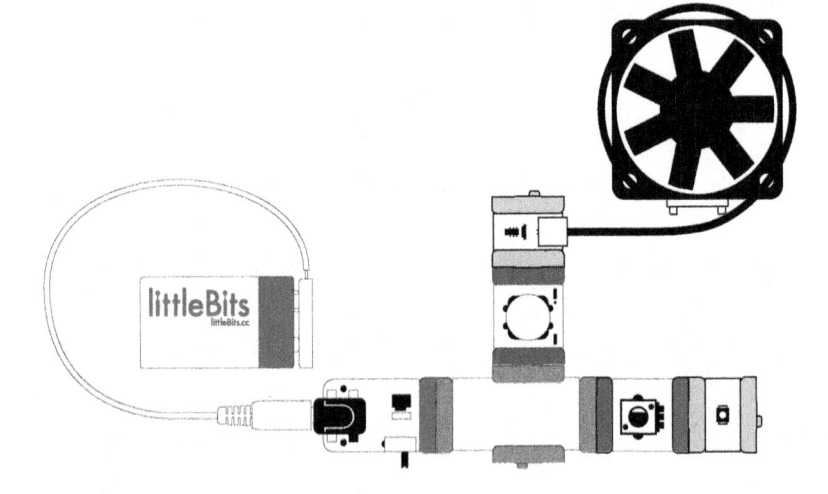

Figure 1-19. *The button will control the fan and the dimmer will control the LED.*

The wire, fork, and branch modules are the simplest wire modules. There are many other wire modules available to you and they're quite powerful. They'll be covered in the chapters ahead!

Troubleshooting Tip

Magnets are magical but they also tend to attract stubborn dust. If your circuit isn't working as expected, try wiping the connectors with a clean, soft cloth—an old T-shirt right out of the laundry is perfect! If that doesn't work at first, carefully wipe all the connectors in the kit—this will often resolve the problem.

Other Accessories

There are a few other accessories that will help you as you work with littleBits.

Screwdriver

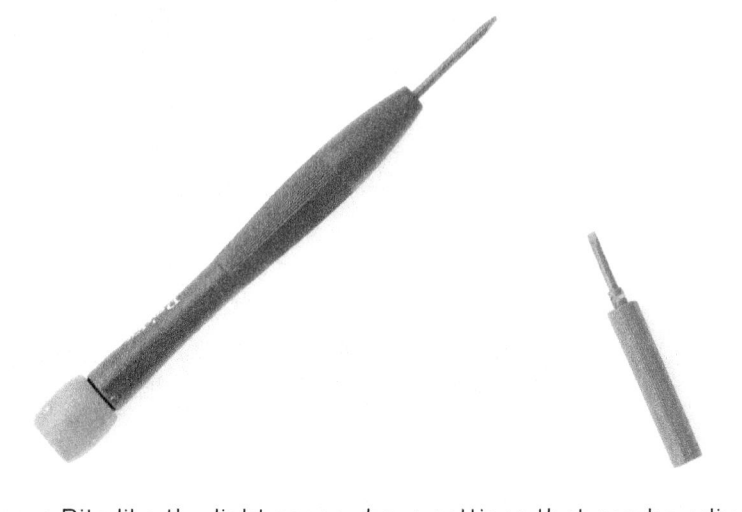

Some Bits like the light sensor have settings that can be adjusted with a 1.8mm flathead screwdriver. Small purple screwdrivers are included with some littleBits bundles for just this purpose. There is a screwdriver included with the Base Kit.

motorMate

Also included in the Base Kit, the littleBits motorMate connects to the DC motor so that you can easily attach wheels, paper, cardboard, and lots of other materials. A Lego axle also fits in the end, so that you can create cool things like the simple drawing bot (*http://littlebits.cc/projects/art-bot—2*) pictured in figure Figure 1-20.

Figure 1-20. *The motorMate helps you connect things to the DC motor, like the Lego axle and wheel you see here.*

Mounting Boards

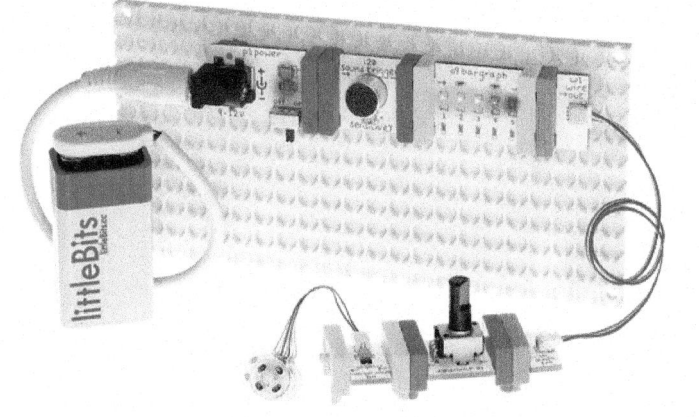

The bitSnap connectors not only connect the Bits to each other, but are also used to snap your project onto mounting boards. These are helpful to add stability and reinforce the magnetic connections you make between Bits.

AC Switch

The AC switch is an IR (infrared)-controlled electrical socket that pairs with the IR transmitter (o18). We discuss the AC switch and how it can be used to build your own smart home devices in Chapter 4.

Shoes

Shoes are another way to lock Bits together by their bitSnap connectors. Connect two Bits, and snap a Shoe on underneath them. Once you connect a Bit to a Shoe, they can also be attached to other materials.

There are three types of shoes: magnetic, adhesive, and hook & loop. Magnetic shoes are great for putting your project on the fridge, or any other material that magnets are attracted to. Adhesive shoes can secure your project to a variety of different surfaces such as plastic, glass, cardboard, and wood, though keep in mind that it's meant for one-time use. Finally, hook & loop shoes are perfect for attaching littleBits to fabric and clothing for wearable projects.

For example, you could make a light-up dog collar, like the one in figure Figure 1-21. Just connect a sound trigger to bargraph modules with a few wire modules. Lock them together with hook & loop shoes. Sew parts of the adhesive strip to the collar, position the circuit, and watch the collar light up when your dog barks. To build one yourself, check out: *http://littlebits.cc/projects/light-up-dog-collar*

Figure 1-21. *This Dog Collar project uses the hook & loop Bits.*

Brick Adapters

The Brick Adapter is another fantastic way to connect Legos and littleBits. One side of the adapter snaps to Lego bricks and the other side snaps to the Bits. If you're connecting two Brick Strips end-to-end, be sure to match the hash marks on the littleBits side.

There are also two types of Brick Adapters: stud and socket. The stud Brick Adapters allow you to push the bottom side of a Lego brick onto your Bits. Socket adapters, on the other hand, allow you snap your Bits onto the top of Lego bricks.

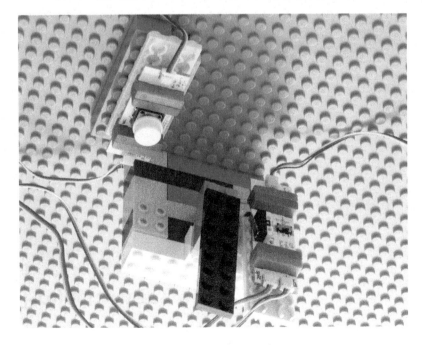

Figure 1-22. *Brick Adapters let you connect littleBits and Lego.*

Project: Night Airplane

http://littlebits.cc/projects/night-airplane

The Night Airplane (*http://littlebits.cc/projects/night-airplane*) project by Konstantin Bauman from Millburn, NJ is perfect for a child's bedroom. This moving mobile hangs from the ceiling and activates when the lights go out at night. You can find all these Bits in the Base Kit and you'll need a few other materials.

Bits used:

- Power
- DC motor
- Light sensor
- Bright LED
- 2x Wire

Other materials, tools, and accessories:

- motorMate
- Screwdriver
- Paper and cardboard

- Tape and string

To make your own:

1. Build out the circuit as shown in Figure 1-23.
2. Set the light sensor trigger switch to dark mode.
3. Using the screwdriver, adjust the sensitivity to match the environment where the Night Airplane will be placed. The motor should run only when it's dark.
4. Create a paper airplane and tape a loop of string to the front and back.
5. Tie the string to one end of a small piece of cardboard and it secure with tape, if necessary. This will act as an arm so that the airplane can fly in loops through the air rather than just spin in place.
6. Place the other end of the piece of cardboard into the slot of the motorMate so it fits snugly. Secure with tape if needed.
7. Mount the circuit to the ceiling with tape or adhesive mounting shoes. Of course, the DC motor will need to be mounted so that the axle faces downward.

✏️ To mount the Bits to the ceiling, use adhesive shoes or tape. You can also place the circuit on a mounting board and tape that to the ceiling.

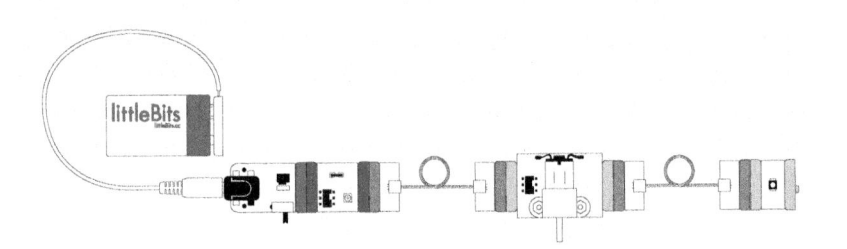

Figure 1-23. *The Night Airplane project circuit.*

Project: Coffee Table Ground Effect Lighting

http://littlebits.cc/projects/coffee-table-ground-effect-lighting

This project, Coffee Table Ground Effect Lighting (*http://little bits.cc/projects/coffee-table-ground-effect-lighting*), is a simple use of inputs and outputs that goes beyond the Bits in the Base Kit.

Parties are so much better with lighting effects, aren't they? Matt was having some guests over, so to help with the ambiance, he added some downward-pointing LEDs underneath his coffee table, which gives it a nice underglow. He also added a slide dimmer on the top of the coffee table so that the effect can be easily adjusted.

Bits used:

- Power
- 5x Bright LED (adjust quantity for the length of your table)
- 6x Wire (adjust quantity for the length of your table)
- Dimmer or Slide Dimmer

Other materials, tools, and accessories:

- Masking tape or adhesive shoes
  - Create the circuit as shown in Figure 1-24.
  - Arrange the Bits across your table to make sure you have the right number of bright LED Bits and wire Bits.
  - Using tape or adhesive shoes, attach the Bits to the underside of the table.

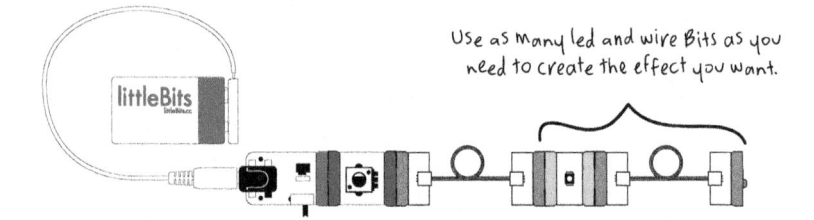

Figure 1-24. *The basic circuit for Coffee Table Ground Effect Lighting. The amount of wire and LED Bits you use will depend on the length of your table.*

A Quick Overview of littleBits Resources

Project Tutorials
  Looking for inspiration? You can peruse thousands of projects, from smart home devices to music to robotics, on littlebits.cc/projects

Tools for Educators
  For lessons plans, case studies, and educational discounts, check out littlebits.cc/education

Troubleshooting
  For help with a specific module, check out the littleBits Forums (*http://discuss.littlebits.cc/*) or check out the Tips & Tricks (*http://littlebits.cc/tips-tricks*).

Library

Remember, the littleBits library is always growing! Check the library (*http://littlebits.cc/shop?filter=Bits*) or the newsletter for updates on new or improved modules.

While littleBits can foster beginner experimentation with electronics, the littleBits library's power is almost limitless. Now that you've got the basics down, the chapters ahead will unlock the full potential of littleBits and show how to make projects with high levels of complexity, but with very little friction.

2/Control and Logic

There's a lot of power in the ability to easily chain inputs and outputs together with littleBits. You can fine-tune that power with a few Bits that give you control over how the signals flow through your project.

These Bits are handy because they control how the signal is passed between the modules. These Bits may seem simple at first, but don't underestimate them. In no time you'll find yourself reaching for them in many (if not all) of your projects!

Let's start by taking a look at a very contrary Bit, the Inverter.

This chapter uses all the Bits in the Logic Expansion Pack (*http://littlebits.cc/expansion-packs/logic*). To try out the examples, you will also need a fork, branch, or split Bit, two wire Bits, two input Bits of your choice, and two output Bits of your choice. This chapter also covers the pulse and timeout Bits, which you can purchase individually, or as part of the Deluxe Kit (*http://littlebits.cc/kits/deluxe-kit*), as well as the threshold Bit, which is included in the Smart Home Kit (*http://littlebits.cc/kits/smart-home-kit*). You can generally make substitutions with input and output Bits as needed, but because of the unique behavior of the pulse and timeout input Bits, you should use those when directed.

If you have the Logic Expansion Pack and the Deluxe Kit, you'll be able to do most of the examples in this chapter (you'll be missing the motion sensor and roller switch Bits).

Inverter

The inverter (w10) is one of the simplest wire modules, but it's very handy. There are cases when you want the output of a Bit to be the opposite: whenever it receives an ON signal, it outputs an OFF signal and vice versa. That's all the inverter Bit does. It outputs the opposite of its input. Simple, right?

Try out the inverter yourself in its simplest form:

1. Connect a button to a power module.
2. Connect an inverter to the button
3. Connect an LED to the button, as shown in Figure 2-1. Notice that when the button isn't pressed, the LED is on. When you press the button, the LED goes off.

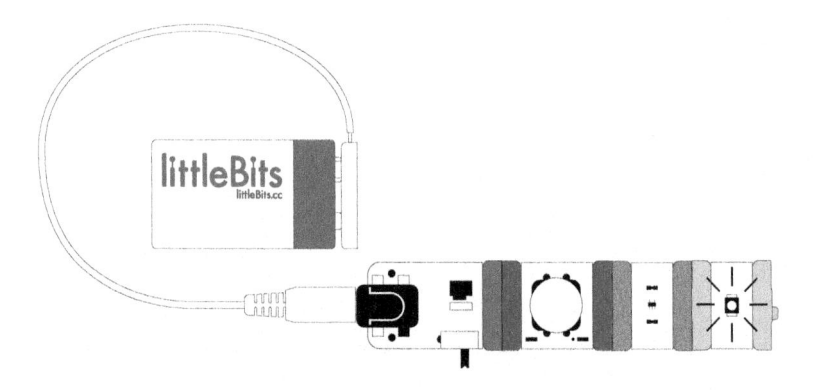

Figure 2-1. *Connecting a button and LED to the inverter Bit.*

Digital and Analog

Try the inverter between an analog input and an analog output, like a dimmer and bargraph module. For the dimmer, you can either use the regular dimmer Bit (i6) from the Base Kit, or the slide dimmer Bit (i5) from the Premium Kit and Deluxe Kit.

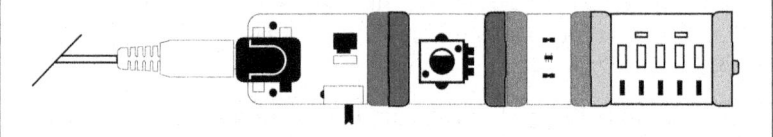

As you adjust the dimmer, you'll notice that the bargraph is only entirely on or entirely off. This is because the inverter only works in a digital manner. There's a threshold around 2.5 volts and the module will turn on or off depending on whether it's above or below this threshold.

If you put an o21 number Bit (in Volts mode) between the power Bit and the inverter Bit, you can discover this voltage for yourself by watching the output of the number Bit.

When the inverter senses an input voltage below the threshold, it outputs an ON signal. Inversely, when the voltage is above the threshold, it outputs an OFF signal.

Pulse

The pulse Bit (i16) toggles the signal ON and OFF repeatedly. You can adjust its speed by turning the small dial with a screwdriver. If you want to make something blink, the pulse Bit is what you should reach for.

At its slowest, the pulse Bit toggles the signal about once per second. When you turn it up all the way, it goes quite fast, almost giving a strobe effect. This strobe effect came in handy for the zoetrope project shown in Figure 2-2. When the 3D printed characters shown in Figure 2-3 spin and the timing of the pulses of light is just right, you'll see the character come to life as an animation! Check out the project page (*http://little bits.cc/projects/zoetrope*) to see it in action or to learn how to build your own.

Figure 2-2. *The outside of the zoetrope project.*

Figure 2-3. *The inside of the zoetrope has 3D printed characters which turn into an animation.*

Project: Flashing Sign

Try out the pulse Bit with the inverter Bit to make a flashing sign, much like the neon signs you'd see in a diner window. Usually they alternate between two words like "EAT" and "HERE" or "OPEN" and "NOW." But yours can alternate between "HAPPY" and "BIRTHDAY" or "WELCOME" and "HOME."

Here are the Bits you'll need:

- 1 power
- 2 LEDs (you could also use bargraph or light wire Bits)
- 1 inverter
- 1 pulse
- 1 wire

Here are the other materials:

- small cardboard box
- paper
- tape
- inkjet or laser printer
    - Connect the Bits as shown in Figure 2-4 and place them inside your box (Figure 2-5). You may need to use a wire Bit depending on where you want to place the LEDs.
    - With a screwdriver, set the pulse Bit to its slowest setting.
    - Using your preferred graphics program on your computer, lay out your message and print it out. It may take some trial and error so that the words will line up with the LED Bits. You can also use additional wire Bits to space out the LEDs if necessary
    - Cut holes in the cardboard box so that they're behind the words.
    - Arrange the LED Bits so that they're behind each word.

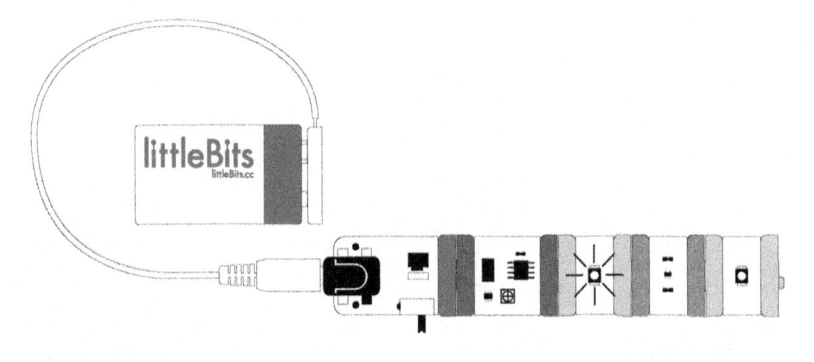

Figure 2-4. *The pulsing ON signal from the pulse module will turn on the first LED and the inverter will turn off the second LED. When the pulse Bit turns the LED off, the inverter turns the second LED on.*

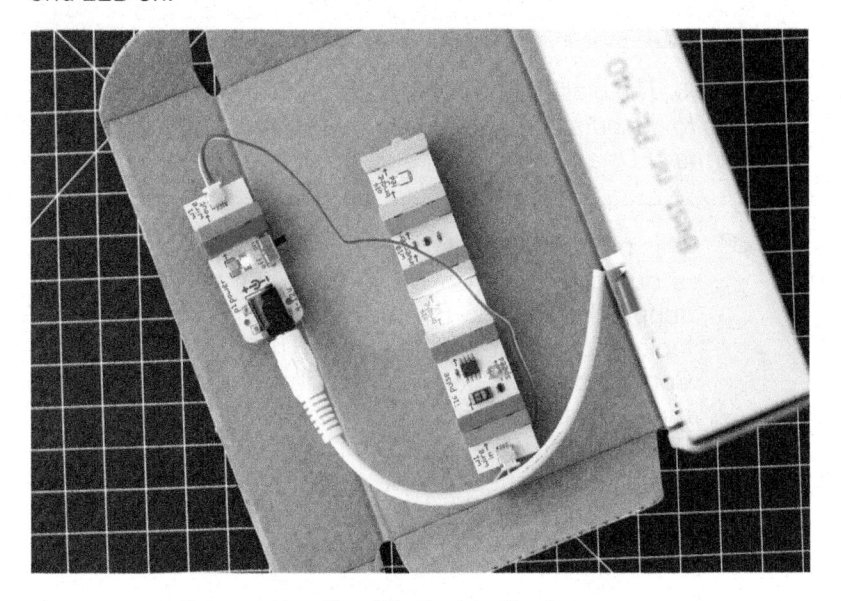

Figure 2-5. *Connecting the Bits inside the box.*

When the pulse is ON, the first LED will light up. For the second LED, when the pulse is ON, the inverter will change it to OFF and vice versa. The effect gives you two flashing LEDs that are alternating, much like those neon diner signs!

Figure 2-6. *The piece of paper with the message printed on it attaches to the outside of the box so that the cut holes are lined up with the LEDs and words.*

This project demonstrates that even if an OFF signal is outputted from a Bit, there's still voltage being supplied to all the Bits in the circuit. If you take a look at the overview in "Under the Hood: bitSnap Connectors" on page 13, you'll see that there are three contacts: ground, voltage, and signal. Input Bits and the Bits covered in this chapter all have some kind of effect on the signal, but as with all Bits, they pass the ground and voltage connections to the Bits that connect to them.

If you can get two LEDs to alternately flash, it means you can also do the same with any other output module. If you have two light wire Bits, you can even make a neon-style diner sign!

Latch

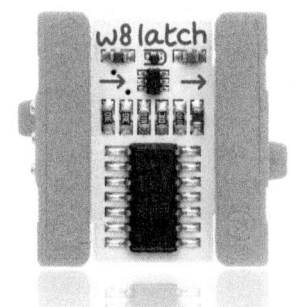

The latch module (w8) acts like a toggle switch, a switch that holds an ON or an OFF signal. If you send it a momentary ON signal (like a single press from a button), it will hold an ON signal as an output. Send another momentary ON signal and the latch's output will flip back to OFF. Essentially, it converts a momentary input into a toggling input.

Try out the latch in its simplest form:

1. Connect a button to a power module.
2. Connect a latch to the button.
3. Connect an LED to the button.

Your circuit should look like Figure 2-7. When you press the button, the LED should turn on and stay on. Press the button again and the LED will turn off and stay off.

Just like with the inverter, the latch only works as a digital Bit. Any input voltage less than around 2.5 volts will be treated as off and 2.5 to 5 volts treated as on. The latch can be helpful as a way of turning an analog input into an on/off switch. For instance, a pressure sensor (i11) or bend sensor (i14) module in conjunction with a latch can be used to turn an output on and off.

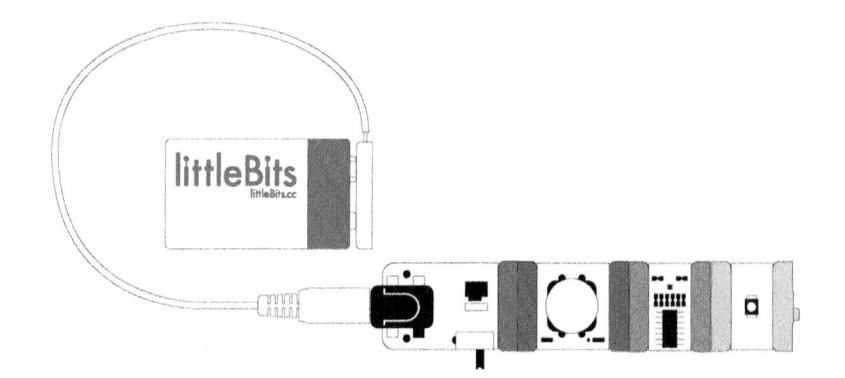

Figure 2-7. *Connecting a button and LED to the latch Bit.*

Geek Speak: Latches

Your computer's memory, or RAM, is actually made of billions of electronic latches. And just like like a latch in your computer's memory, the latch Bit can store information as long as it has power. One latch Bit is the equivalent of a single computer Bit and can represent either ON or OFF (1 or 0 in *binary*). You'd need 8 of these Bits to store a single byte. With all the different combinations of ON and OFF between the 8 Bits in a byte, you can store the values 0 through 255. On the littleBits site, you can learn how to make your own electronic memory using latches! (*http://littlebits.cc/fridays-tips-tricks-latch*)

Timeout

If you have a momentary ON signal and you need to extend it to be a longer signal, the timeout Bit will come in handy. For example, let's say you want to create a motion-activated intruder alarm. When the timeout Bit receives an ON signal from the motion sensor, it can keep the buzzer going for a few minutes and then turn it off. With the included screwdriver, you can adjust the timeout to be really short, less than a second, or really long—up to about 5 minutes, plenty of time to annoy an intruder!

There's also a mode switch on the timeout Bit. In "on-off" mode, the timeout Bit will hold an ON signal for a period of time after it receives an ON signal. In "off-on" mode, the timeout hold an OFF signal when the Bit receives an ON signal, otherwise, it keeps the signal ON.

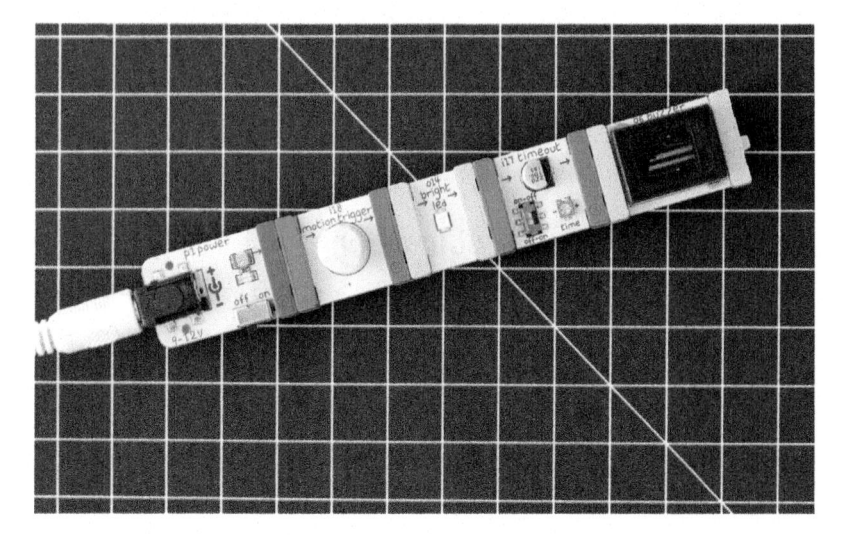

Figure 2-8. *This simple intruder alarm catches motion and activates a buzzer, even after the motion has stopped.*

Try out the timeout Bit to make a simple intruder alarm:

You will also need a motion trigger Bit for this project, which is not included in the Deluxe Kit. If you don't have a motion trigger, you could build this project using the sound trigger Bit (i20, which is included in the Deluxe Kit) in place of the motion trigger Bit. If you do this, be sure to use an LED as your output rather than the buzzer so that the alarm doesn't trigger itself over and over! If you use the sound trigger bit, you'll only catch noisy thieves with this.

1. Connect a motion trigger Bit (i18) to a power Bit and then connect a buzzer (o6) directly to the motion sensor so that you can see how the motion trigger behaves. When the motion trigger senses motion around it, it will send an ON signal until a few seconds after things around it get still.

2. Add the timeout Bit between the motion trigger and the buzzer. Make sure that the timeout Bit is in ON-OFF mode.

3. Using a screwdriver, adjust the time setting on the timeout so that the buzzer is still buzzing for a few seconds after there's stillness.

4. To help see the behavior of your circuit, also try adding an LED Bit between the motion trigger and the timeout. This way, you'll get an indication of when the motion trigger no longer senses motion even if the timeout Bit is still sending the ON signal to the buzzer.

Your circuit will look something like Figure 2-9.

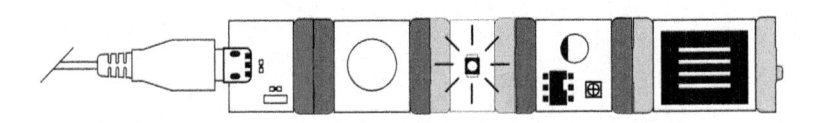

Figure 2-9. *The buzzer in this circuit will continue to buzz for a minute after the motion is first sensed. The LED indicates when the motion trigger is passing the ON signal to the timeout Bit.*

Project: Midnight Snack Light

http://littlebits.cc/projects/midnight-snack-light

When you're grabbing a midnight snack, the light inside your refrigerator is so convenient when the door is open, but after you've closed the door, you're in the dark again. The midnight snack light not only lights up the counter next to the fridge when you open the door, but keeps it on for a couple minutes after you've closed it so that you can see what you're eating.

Here are the Bits, accessories, and materials you'll need:

- One or more LED Bits (use three bright LED Bits for maximum effect)
- 1 roller switch Bit (included in the Premium kit but not the Deluxe Kit). You should be able to improvise something with the regular button Bit if you don't have the roller switch.
- 1 timeout Bit
- wire Bits (amount depends on your setup)
- power Bit. We recommend the USB power Bit with its USB power adapter and micro USB cable
- littleBits Mounting Board (optional, but recommended)
- 3M Command strips (optional, but recommended)
- A piece of plastic to trigger the roller switch (use some scrap plastic, mold your own with sculptable plastic like ShapeLock, or 3D print it!).
- Screwdriver

To make it:

1. Assemble the circuit as shown in Figure 2-10, placing wire modules where needed. Set the roller switch to "open" mode.

2. Attach the roller switch to the side of the refrigerator using a mounting board with 3M Command Strips on the back.

3. Attach a piece of plastic to the side of the door of the fridge so that it pushes the roller switch closed when the door is closed. If you have a 3D printer, you can print your own from this design: *http://www.thingiverse.com/thing:482092* (see Figure 2-11)

4. Use a screwdriver to adjust the timeout's interval.

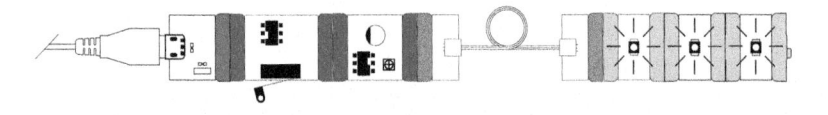

Figure 2-10. *The basic circuit for the Midnight Snack Light*

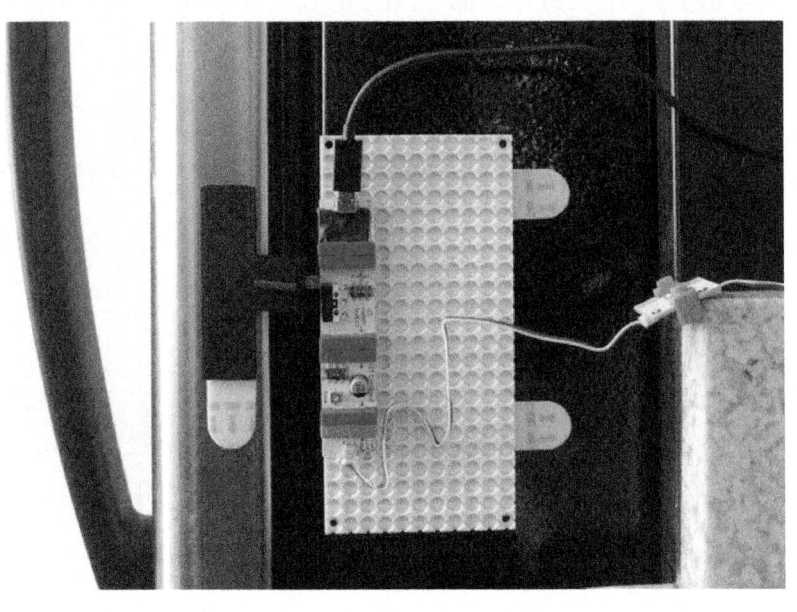

Figure 2-11. *Using 3M Command Strips to mount the mounting board and triggering bracket to the refrigerator.*

Threshold

The threshold (i23) compares the signal coming into the module's input connector to a voltage set by the knob. If the input voltage is greater than the selected voltage, the output is set to max voltage (5V). Use it to make any sensor module into a trigger module. You can combine the threshold with the temperature sensor (i12) and a number Bit (o21) to create an alarm that triggers when your fridge gets too warm!

Try out the threshold with the following:

1. Connect a bend sensor to a power module.
2. Connect a threshold to the bend sensor.
3. Connect a bright LED to the threshold.
4. Set the voltage by turning the threshold knob.
5. When the bend sensor reaches the threshold level, the bright LED will turn on!

Logic Bits

The littleBits logic modules help you create rules for your circuit to follow, which opens up the possibility for more complex circuits. In electronics, these components are called *logic gates*. For instance, the intruder alarm from "Timeout" on page 43 could be set up with logic (and the addition of the sound trigger Bit) so that if it detects motion **or** a loud noise, the buzzer will turn on.

 You've actually already tried out a logic module earlier in this chapter: the inverter, which simply reverses the logic that it receives.

If you've been following the examples in this book up until now, you've mostly been making *series* circuits that follow a single path and perhaps branch into multiple outputs. What happens when you have multiple *inputs* and you want them to affect a single output? This is where the logic Bits are useful.

 Like with the other Bits you've read about in this chapter, the logic Bits work in the realm of digital, that is, only on or off.

Before we look at each logic Bit, set up a circuit as shown in Figure 2-12.

In addition to the logic modules introduced in the following sections, you can try these logic Bit experiments out with these Bits:

- power
- fork, branch, or split
- 2x input Bits such as slide switch or button
- 3x output Bits such as RGB LED, LED, or bargraph
- 2x wire

You don't need to follow the exact circuit in Figure 2-12. For example, swap the toggle switches for buttons or use another type of output. The important thing is that you can control two separate outputs, connect them both to a logic Bit, and connect an output to the logic Bit.

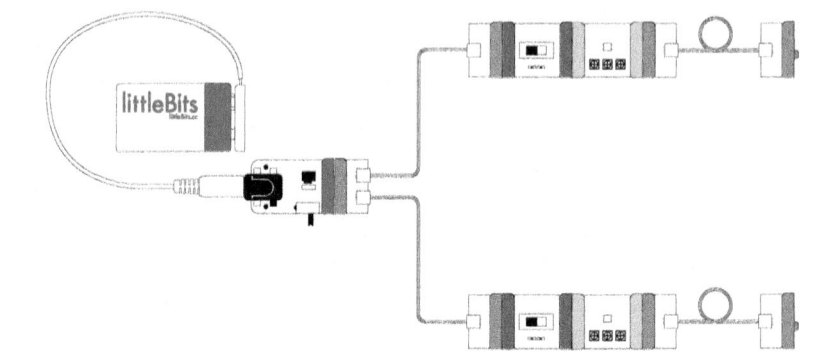

Figure 2-12. *To try out the logic Bits, set up a fork, branch, or split with two input Bits.*

Double AND

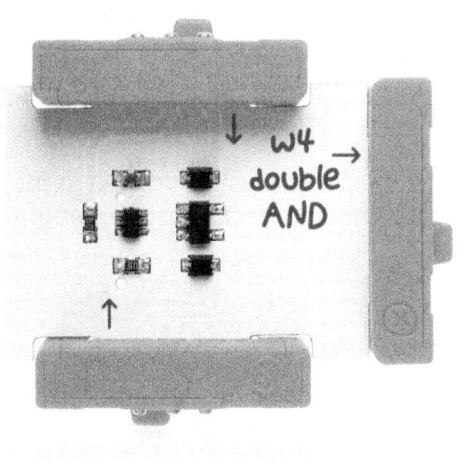

With a double AND Bit (w4), both input signals must be ON in order for it to output an ON signal.

Connect it to the wires as shown in Figure 2-13 and add an output Bit (like the bargraph) to the output side of the double AND Bit. Try out the two switches and you'll see for yourself that both of them must be turned on in order to turn on the bargaph Bit.

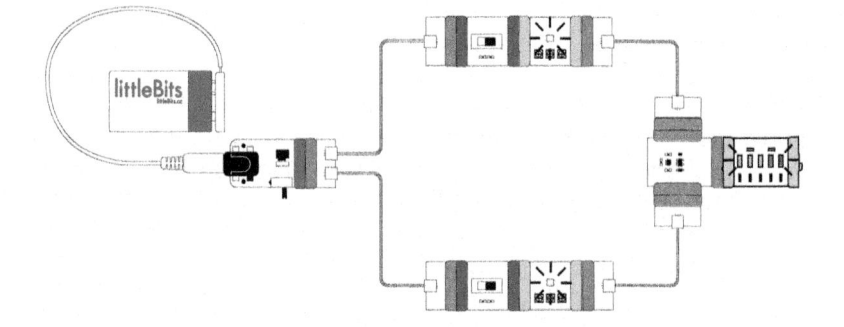

Figure 2-13. *Only when both of the switches are on is the bar-graph module turned on.*

Here's the *truth table* (the output that corresponds to each *input state*) for the double AND Bit:

Input 1	Input 2	Double AND Output
OFF	OFF	OFF
ON	ON	**ON**
OFF	ON	OFF
ON	OFF	OFF

The littleBits team used the double AND Bit in their Morning Sunshine (*http://littlebits.cc/projects/morning-sunshine*) project as shown in Figure 2-14. It's an alarm that only goes off when the sun is bright enough and you're still in bed. The two inputs are a light trigger for sensing the sun and a roller switch, which is ON when your head is against the pillow.

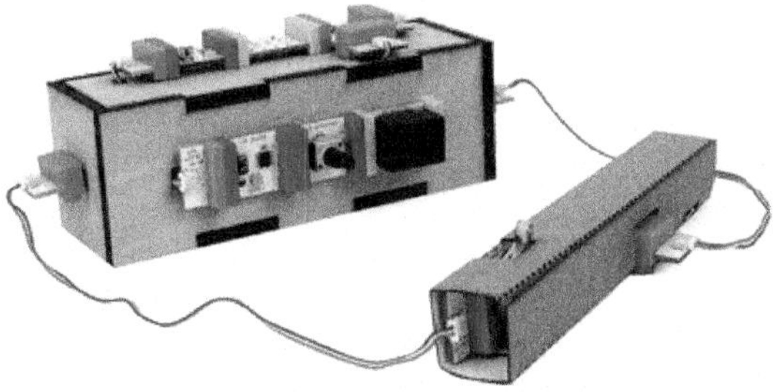

Figure 2-14. *The Morning Sunshine project*

Double OR

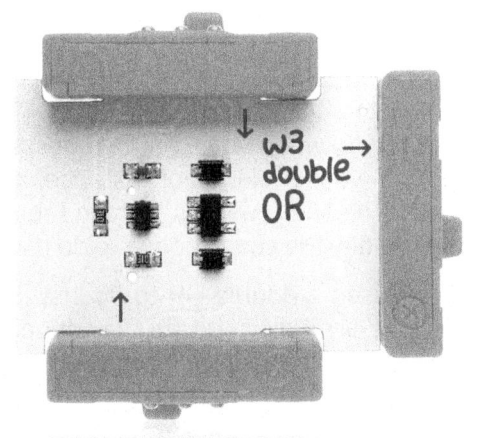

Now swap the double AND Bit for a Double OR Bit (w3), as shown in Figure 2-15.

With a double OR, when either input is ON, it will output ON. It's also important to note with the double OR that when both inputs are ON, it outputs an ON signal.

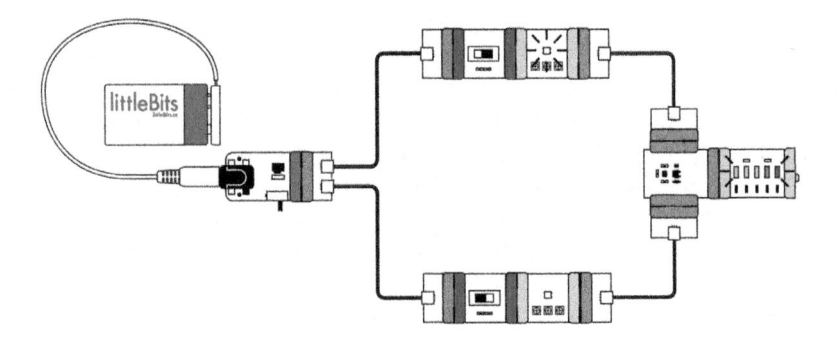

Figure 2-15. *When either switch is ON, the double OR outputs an ON. When both are ON, it also outputs ON.*

Here's the truth table for the double OR bit:

Input 1	Input 2	Double OR Output
OFF	OFF	OFF
ON	ON	**ON**
OFF	ON	**ON**
ON	OFF	**ON**

The double OR is useful when you want two possible ways for a user to interact with your project. For instance, a slot machine may have a button to send the reels spinning, but if the player wants to pull the handle, they can choose to do that instead.

The littleBits team used a double OR to do just that when they made their own one-armed bandit for Valentines Day (*http:// littlebits.cc/projects/littlebits-lucky-slot-machine*).

NAND

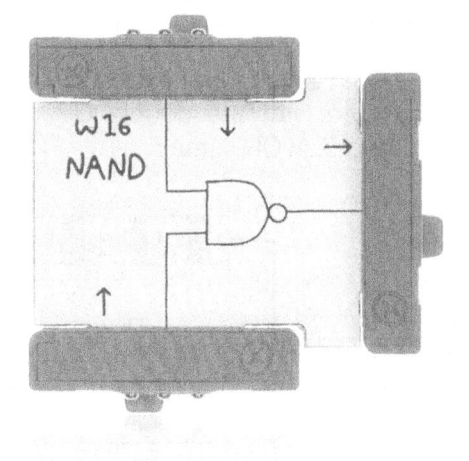

Now remove the Double OR and put the NAND Bit (w16) in its place, as shown in Figure 2-16. NAND means "NOT AND." It does the exact opposite of the double AND. When both inputs are ON, it outputs OFF. In any other case, it outputs ON.

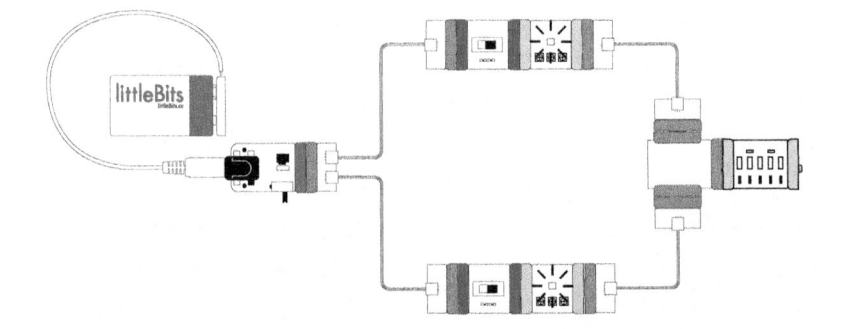

Figure 2-16. *The only case when the NAND Bit outputs an OFF signal is when it receives two ON signals.*

Here's the truth table for the NAND Bit:

Input 1	Input 2	NAND Output
OFF	OFF	**ON**
ON	ON	OFF
OFF	ON	**ON**
ON	OFF	**ON**

You can use two NAND Bits to create a two player game show lockout buzzer. The idea is that when a player presses their button, it disables the other player's button. That way, there's never a dispute about who rang-in first. Figure 2-17 shows a how a game show lockout circuit would look.

Until one of the players presses their button, both NAND gates are sending a signal to the inverter, which keeps the LED off (because it's inverting the signal). At the same time, each NAND gate sends a signal also to the other NAND gate (because the branch Bit splits its signal). Until someone presses a button, each NAND gate is getting only one input (a signal from the other NAND gate).

The first player to press their button causes the NAND gate to receive signals from both the button and the opposite NAND gate. This causes it to stop sending a signal, which the inverter inverts, illuminating the LED.

Now what happens when the other player presses their button? At this time, there is no signal coming from the other NAND gate, nor will there be (until the other player lifts their hand from their button). This means that it will always send a signal to the inverter, whether the other player presses their button or not. And the inverter reverses that signal, keeping the LED dark.

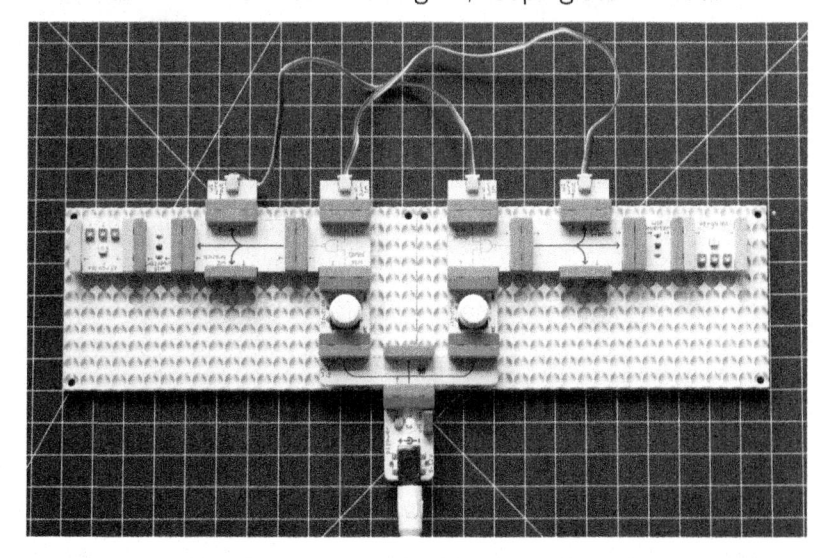

Figure 2-17. *With two NAND Bits, you can make a game show lockout system.*

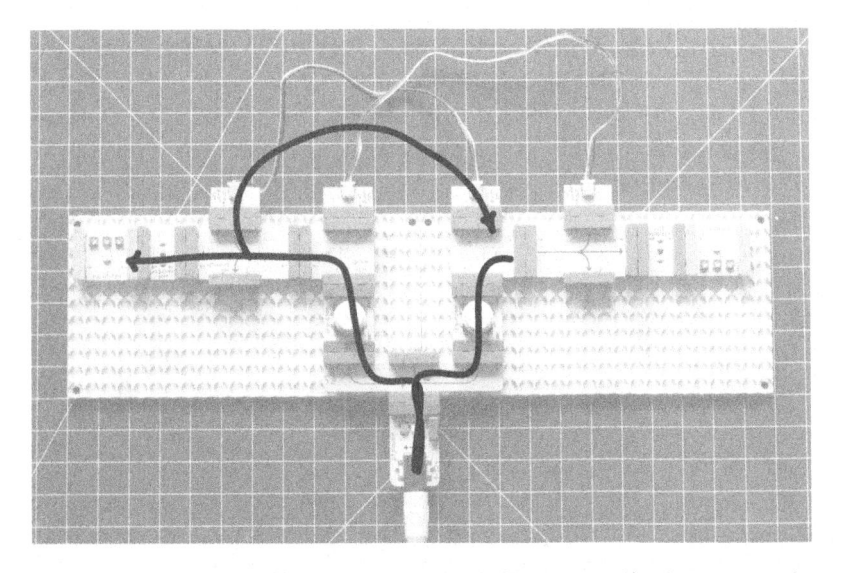

Figure 2-18. *When the player on the left presses the button and turns their LED on, it disables the player on the right from turning their LED on.*

NOR

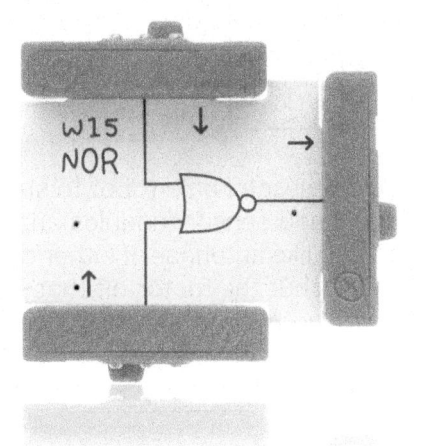

As you may have guessed, the NOR Bit (w15) does the opposite of the Double OR Bit. In other words, when both of the inputs are OFF, then the NOR Bit outputs an ON signal. In any other case, it outputs OFF. Try it out yourself, building the circuit as shown in Figure 2-19.

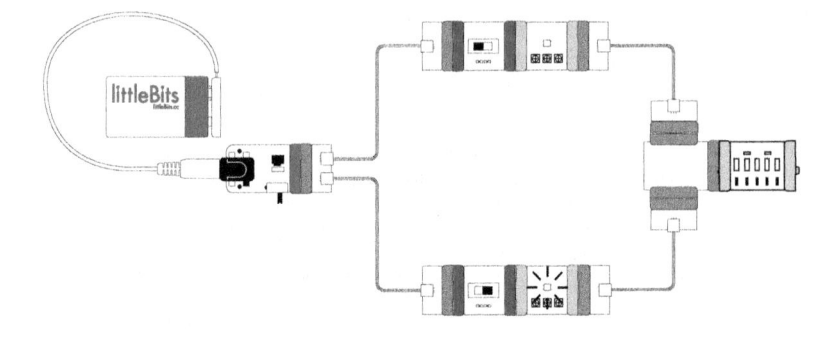

Figure 2-19. *When both inputs into the NOR Bit are OFF, it outputs ON.*

Here's the truth table for the NOR Bit:

Input 1	Input 2	NOR Output
OFF	OFF	**ON**
ON	ON	OFF
OFF	ON	OFF
ON	OFF	OFF

The NOR Bit would be useful on a robot to stop it from running into walls. Use two bend sensor modules and have them point out in front of the 'bot like antennae. If either of them bend when they hit a wall, it will shut the motor off that's driving the robot forward. Only when both bend sensors are straight does the robot know that it's safe to drive forward.

XOR

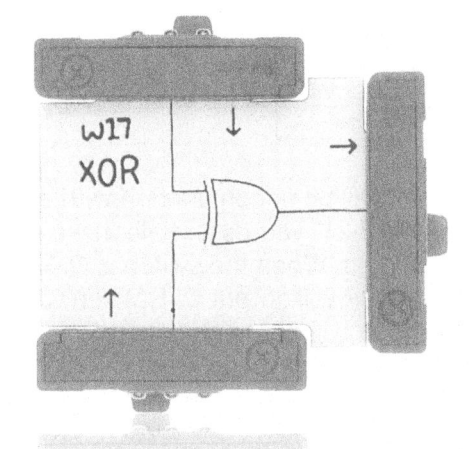

The final logic Bit you'll try in this chapter is the XOR Bit (w17), which means eXclusive OR. If either input is ON, it will output ON. But if both are ON, it will output OFF. If both inputs are OFF, it will output OFF. In other words, it will only output ON when it receives a differing signal from its two inputs. Try it out yourself, as shown in Figure 2-20.

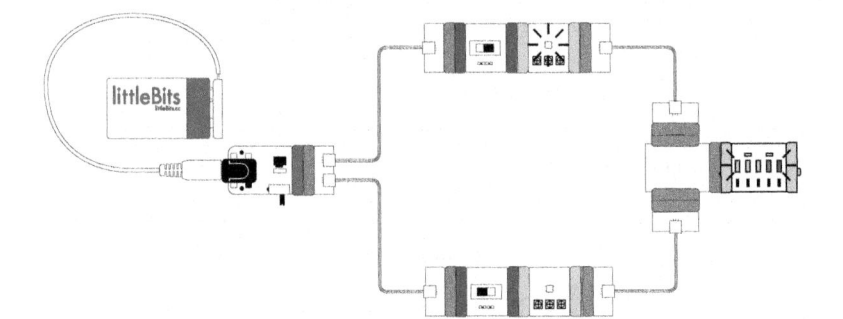

Figure 2-20. *Both inputs have to be different from each other in order for the XOR to output an ON signal. It's the eXclusive OR Bit.*

Here's the truth table:

Input 1	Input 2	XOR Output
OFF	OFF	OFF
ON	ON	OFF
OFF	ON	**ON**
ON	OFF	**ON**

When you use the XOR bit in conjunction with a toggling switch as shown in Figure 2-21, you essentially have an inverter that you can turn on and off. When the switch is OFF, and the button is not pressed, the XOR will output OFF. When you press the button, the XOR output will turn ON and the LED will turn on. Switch the switch to ON and you'll get the opposite behavior: the LED will remain on unless you press the button.

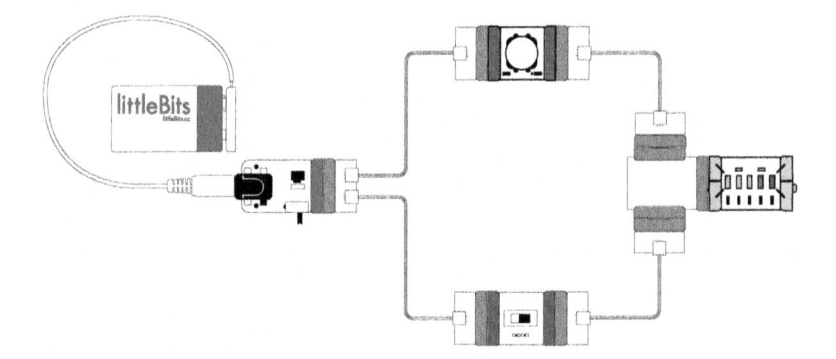

Figure 2-21. *Using the XOR bit to make an inverter that can be turned on and off.*

Also, if you have the littleBits Synth Kit, you can use the XOR to modulate between two oscillators. The result is an eerie, wobbly sound. After you learn about the Synth Kit in Chapter 3, check out the Ring Modulation (*http://littlebits.cc/projects/ring-modulation*) tutorial online.

Going Further

If you want to learn more about what you can do with logic Bits, check out the lessons on the littleBits site. (*http://littlebits.cc/browse-lessons/introduction-to-logic*)

3/Music and Motion

With littleBits, you can make music *and* make things move. This chapter starts with the Synth Kit Bits, which enable you to build your own analog synthesizer. Then it shows you how to work with the motors in the library to give locomotion to your project.

Synth Kit

You might already be familiar with Korg. They're a Japanese company that has been making electronic musical instruments since 1963. In 2013, Korg formed a partnership with littleBits to create the Synth Kit, a set of Bits designed to allow you to make your own electronic instruments. Even if you have no experience with music, you'll find that the Synth Kit makes it easy to create unique sounds and music to enjoy on your own, to share with others, or to use in conjunction with your other projects.

Your musical projects will always use at least two Synth Bits, one to *generate* the audio and the other to *output* the audio: the oscillator and random Bit are two Bits that are capable of generating an audio signal. The speaker Bit outputs that signal so you can hear it. There are a bunch of other Bits in the Synth Kit that let you modify the sound in many unique ways. Bits from other kits will also come in handy to help you make musical instruments.

Oscillator

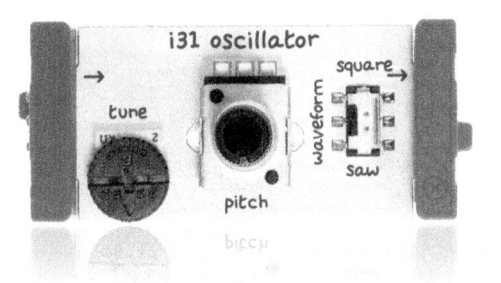

The oscillator (i31) is a source of audio signals in your littleBits instrument. It has adjustments for the tune, pitch, and waveform. In order to understand what each of these characteristics mean, let's first explore the fundamentals of a sound.

Sound is the vibration of air or another medium (like water). When you speak, sing, or clap, you create sound waves that radiate out into the environment. Every sound has its own *signa-*

ture that is called a *waveform*. Take a look at the parts of a waveform in Figure 3-1.

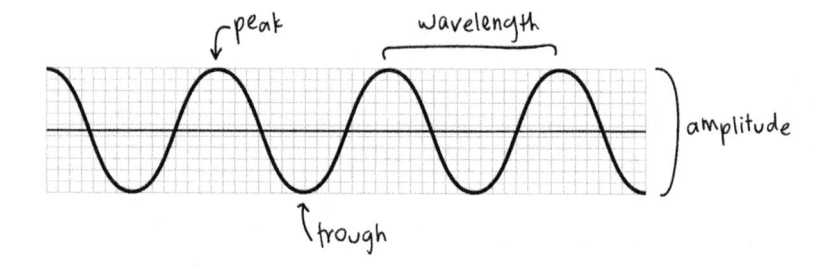

Figure 3-1. *The parts of a sound wave.*

The *pitch* of a sound is related to the *frequency* of its wave: that's how often it makes a complete cycle (one that includes a complete peak and a complete trough). When a sound is at a low pitch, you get a low note... think of a deep cello note. When sound is at a high pitch, you get a high note, like a high note from a flute.

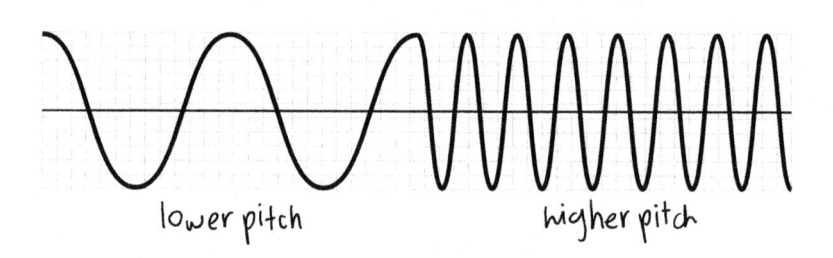

Figure 3-2. *The pitch of the sound is related to the frequency of the wave.*

Frequency and pitch are similar, but not exactly the same. Frequency can be measured scientifically, while pitch is dependant individual perception.

Amplitude relates to the change in the peaks of waveforms and is perceived as the loudness of a sound. The higher the amplitude, the louder it sounds. (See Figure 3-3).

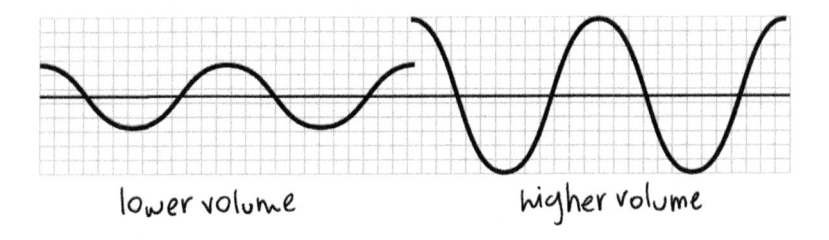

lower volume   higher volume

Figure 3-3. *These two waves have a difference in amplitude, or loudness.*

Try generating a few different pitches now with the oscillator:

1. Connect the oscillator to a power module.
2. Connect a speaker (o22) to the oscillator, as shown in Figure 3-4.
3. Turn the pitch dial on the oscillator.

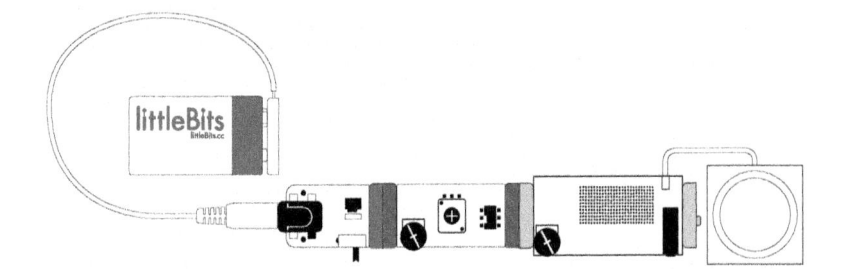

Figure 3-4. *Sweeping the pitch dial on the oscillator lets you hear the full range of pitches that it's capable of.*

The speaker Bit works the way you'd expect: it has a volume control and a headphone jack. Plugging headphones into the jack speaker Bit will disable its on-board speaker so that you can jam without bugging your neighbors.

You should hear the difference between low pitch and high pitch quite clearly. The oscillator's pitch setting even goes so low that it becomes a series of clicks. When it doesn't sound like a tone anymore, it's considered *unpitched*.

Now try changing the waveform switch between square and saw. Hear the difference? The square waveform has a rich, powerful character, and the saw waveform has a more mellow, rounder character. Figure 3-5 shows what these two types of waves look like.

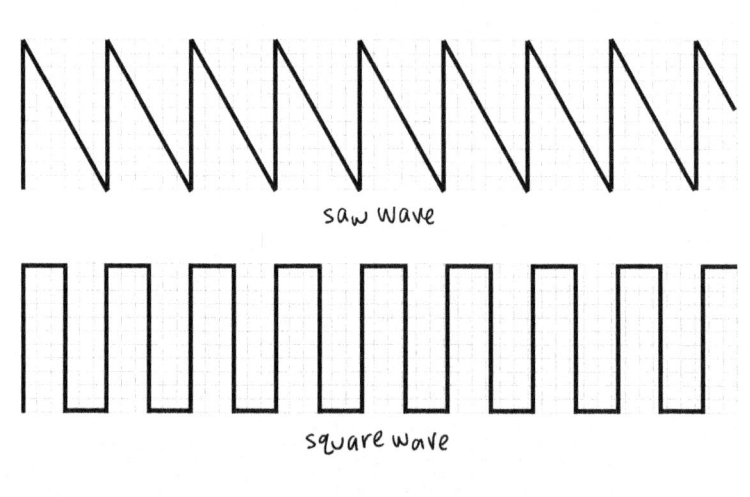

saw wave

square wave

Figure 3-5. *The oscillator Bit can generate saw waves (top) or square waves (bottom).*

If you want to generate some really funky "alien" sounds, connect two oscillator modules together and adjust the pitch of both. Be sure to listen to the effect of those low frequency waveforms in this setup! The results are pretty neat. Using two oscillators like this is called *frequency modulation synthesis*.

Random

Another Synth Bit that generates an audio signal is the random module (i34). The switch on it lets you change between two modes: in noise mode, it generates white noise, just like the sound you hear when an analog radio or TV is not tuned to a broadcast. In random voltage mode, it outputs random voltage signals that can control oscillators and make them play random pitches.

Here are a few things to try with the Random module:

1. Connect the random module in noise mode between a power module and a speaker to hear the white noise. It may not sound great right now, but when you add some bits to it later, you'll see this is really useful.

2. Switch the random Bit to random voltage mode and add a button between the power module and random module.

3. Add the oscillator between the random Bit and the speaker as shown in figure Figure 3-6.

4. Every time you push the button, you'll get a different note.

5. Also try swapping the button for a pulse Bit or a micro sequencer. When you set the pitch of the oscillator and the

speed of the pulse just right, it sounds like the sound effect of an old robot or computer making a calculation!

Figure 3-6. *With the random Bit in random voltage mode, every push of the button translates to a different pitch from the oscillator.*

Keyboard

Now try connecting the keyboard Bit (i30) between a power Bit and an oscillator as shown in figure Figure 3-7. As you may have guessed, the buttons on the keyboard Bit are laid out like a piano. Each key press creates a distinct voltage that represents a note so that you can play melodies. There are 52 different notes that you can play with the keyboard, but only 13 buttons. Use the octave dial to change the range of notes that 13 keys represent. There are 4 different octaves.

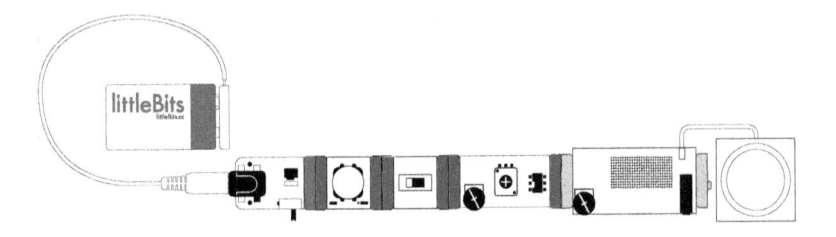

Figure 3-7. *Each button on the keyboard sends a different voltage to the oscillator, which translates them into pitches. This will let you play melodies.*

The key mode switch sets how the keyboard behaves when you press a button. In press mode, it will only play a note while the key is pressed. In hold mode, it will sustain that note until another key is pressed.

Tuning

Tuning is the relationship between the pitches in a musical instrument. Like any other musical instrument, synthesizers need to be tuned so that you can play recognizable melodies. This becomes important if you want to use the keyboard to play music. To tune your littleBits synth project, you'll adjust the tune dial on the oscillator. Here's how:

1. Hold a key on the keyboard and adjust the octave dial so that the tone is in the middle range.
2. Play all the notes on the bottom from of the keyboard consecutively from left to right. This is called a *major scale* in music. You may recognize it as do-re-mi-fa-so-la-ti-do.
3. Turn the pitch knob on the oscillator to adjust the frequency.
4. Play do-re-mi again. If the notes don't quite sound right, try slowly adjusting the tune dial counter-clockwise until it sounds "in tune."

Playing a Melody

Try playing a melody. Figure 3-8 shows you how to play "When the Saints Go Marching In."

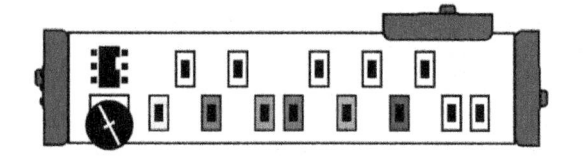

When the saints go marching in

Figure 3-8. *How to play the melody "When the Saints Go Marching In."*

If the keyboard module generates different voltages and those become different notes from the oscillator, this means that any analog input Bit can be used to generate a range of pitches. Try out a dimmer, pressure sensor, or a light sensor between the power module and oscillator to see how they let you control the sound!

Micro Sequencer

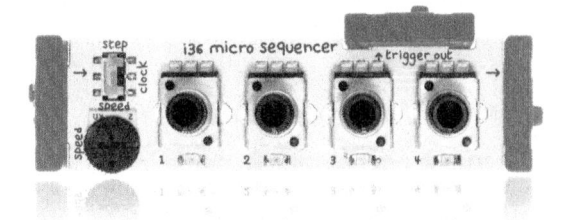

The micro sequencer module (i36) will help you create repeatable melodies when you're making music. It sends out a pattern of voltages based on the position of the four step knobs. In speed mode, you can control the speed of the sequence with the speed dial. In step mode, you can control the sequence with another Bit like a button or pulse module placed before it. Give it a try now:

1. Set the micro sequencer to speed mode and place it as shown in figure Figure 3-9.

2. Play with each of the four step dials and the speed dial to set your melody.

3. Try also silencing some of the steps to by turning it fully counter clockwise.

4. Remove the oscillator and place a random module in noise mode after the sequencer as shown in figure Figure 3-10. Play with the speed and step knobs to create a drum beat.

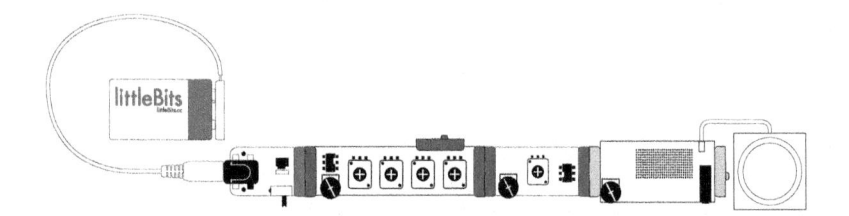

Figure 3-9. *The micro sequencer can send a pattern of voltages that the oscillator will turn into a repeatable melody.*

You've probably noticed that the micro sequencer has a third bitSnap connector labeled "trigger out." Try connecting an LED to it to see its behavior. As you'll see, it pulses the signal along with each step of the sequencer. Maybe you use it to have light wires light up in rhythm with your music. Or try connecting another random module in random voltage mode and an oscillator to the micro sequencer's trigger, as shown in figure Figure 3-10. This will generate random notes for each beat.

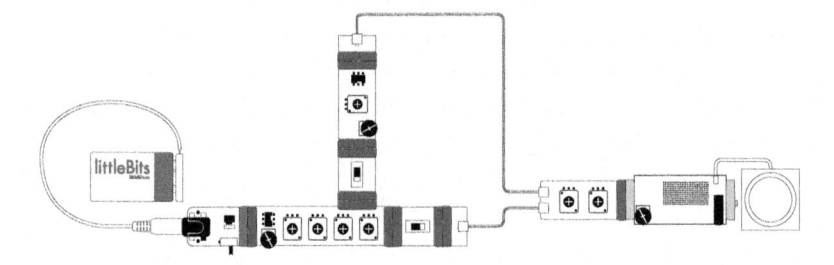

Figure 3-10. *Arranging the Bits like this will create a drum beat and a melody made from random pitches!*

Sequencer

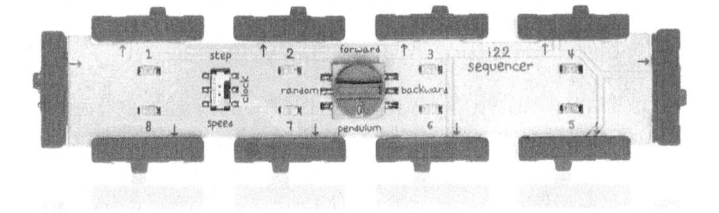

The sequencer allows you to connect up to eight outputs (labeled 1 through 8) and control them in sequential patterns. In "step" mode, the sequence will advance every time the module receives a high signal at the input, like pressing a button. In "speed" mode, the sequence will advance at a speed determined by the input signal. Try using a dimmer to control the speed. It also features a four-position switch that lets you select the direction the sequence runs.

Mix

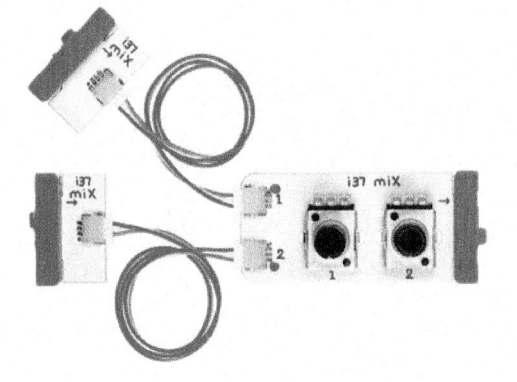

Figure 3-10 showed you the mix Bit (i37) in action. It gives you the ability to take two audio signals and combine them into one. Now that you're branching your instrument into multiple signal paths, it will be useful to control the amplitude, or volume, of each signal when you bring them all together. Each knob on the mix Bit controls the amplitude of one of the incoming audio signals.

The mix Bit can also be used in non-synth projects to combine two input voltages into one.

The split, fork, and branch Bits can be used to split up a single signal into multiple signals if you'd like to put different effects on them and later combine them back together with the mix Module.

Now let's take a look at a few different ways to modify the audio signal.

Envelope

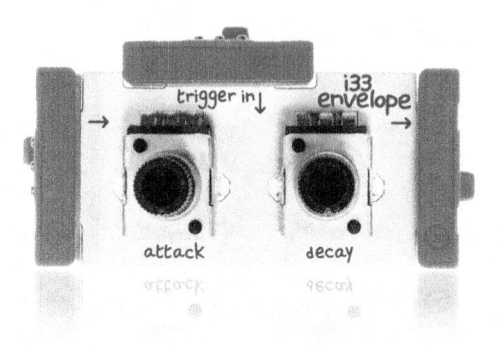

The envelope Bit (i33) affects the amplitude of the signal in two ways: *attack* and *decay*. Attack describes how long it takes to get to maximum volume. Decay describes how long it fades to silence again. Both of these are adjustable with the knobs on board. Try it with the keyboard in hold mode and oscillator as shown in Figure 3-11.

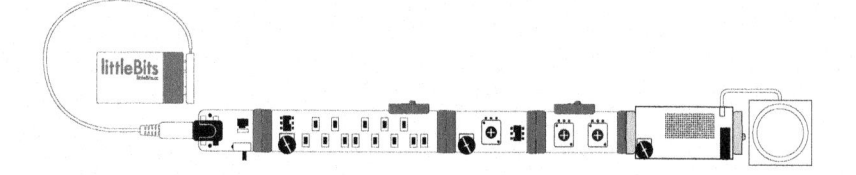

Figure 3-11. *Try out the effect of the envelope Bit by attaching it after a keyboard Bit (in hold mode) and an oscillator.*

A short attack with a long decay is similar to hitting a note on a piano and keeping the key pressed. A long attack with a long decay is similar to bowing a note on a violin. Swap out the keyboard with a micro sequencer to adjust the length of the notes it plays.

If you have a steady audio signal going into the envelope Bit (for example, from a random Bit in noise mode) you can trigger the attack and decay by sending a pulse to the envelope's third bitSnap connector, labeled "trigger in."

Filter

The filter module (i32) has a big effect on the character of the sound, or its *timbre*, by changing the relative volume of certain frequencies in that sound. For example, when you add more high frequencies, the sound will feel "brighter." When you add more low frequencies, it will feel "darker."

The filter module is considered a *low-pass filter*, one that reduces or filters out frequencies higher than a particular frequency.

The filter module has two knobs. The knob labeled *cutoff* sets the frequency that will be emphasized and the knob labeled *peak* sets the intensity of the filter.

You can also use any analog input Bit to adjust the cutoff frequency. That's what the bitSnap connector labeled "freq in" on the filter is for! Try adding a light sensor or slide dimmer Bit to your synth circuit to control the timbre!

Delay

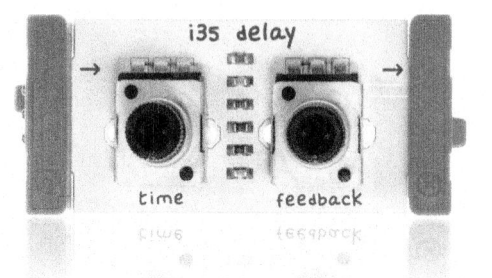

The delay Bit (i35) adds an echo effect to the incoming audio signal. There are two knobs that adjust *time* and *feedback*. Time sets the amount of time between each repetition and feedback sets how many times the sound repeats. Try it with the keyboard Bit and the oscillator as shown in Figure 3-12 to make some futuristic spacey music. Also try making some spooky sounds with the random Bit along with the filter and delay Bits, as shown in Figure 3-13.

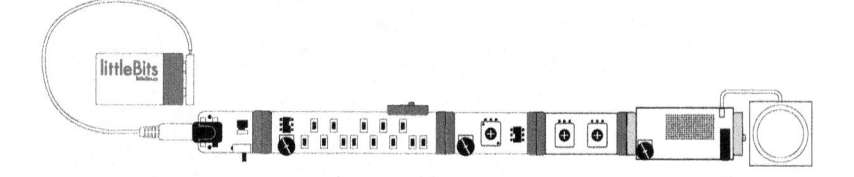

Figure 3-12. *Use the delay Bit to add an echo to your melodies.*

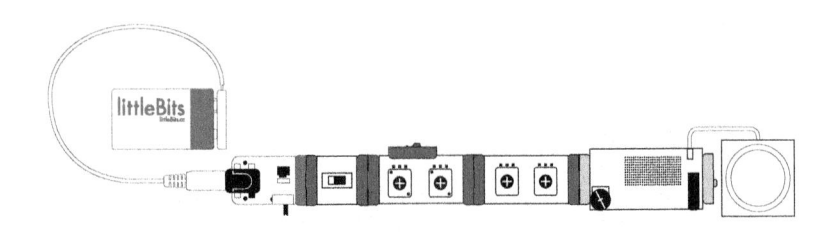

Figure 3-13. *Along with the filter and random Bits, the delay can create some spooky sound effects.*

Project: Synthesizer With the Works

Bring together all your Synth bits to make feature-packed synthesizer! Follow Figure 3-14 and make adjustments to each of the bits until you get just the sound you want from it. You can also swap the keyboard for the sequencer Bit.

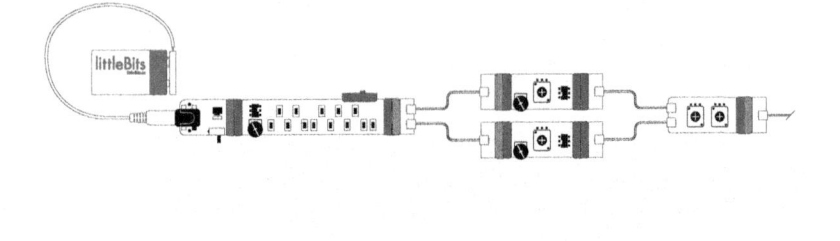

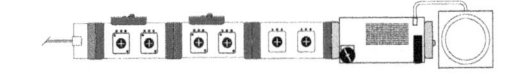

Figure 3-14. *Make a feature-rich synthesizer with many of the synth Bits.*

MP3 Player

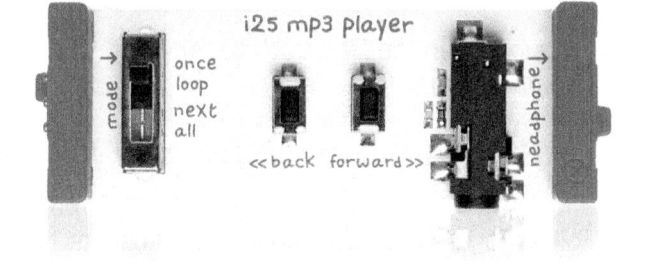

The MP3 Player is found in the Smart Home Kit (*http://little bits.cc/kits/smart-home-kit*) and allows you to play your very own music or sound effects from MP3 files using littleBits. Just load your MP3 files onto the included microSD card. Sending a signal to the MP3 player can make it work as an audio player or sampler. There are two volume levels; press both the forward and back buttons simultaneously to toggle between them.

There is a detailed audio guide loaded on the microSD card.

New from KORG: MIDI, CV, USB i/o

These three modules from KORG make a great addition to the Synth Kit (*http://littlebits.cc/kits/synth-kit*).

The MIDI module (w5) allows you to control the Synth Kit from a digital audio workstation or DAW (e.g., Ableton Live, Pro Tools, etc.) and other MIDI-enabled instruments. Additionally, it will allow you to create your own MIDI controller with littleBits modules by converting littleBits control voltages to MIDI messages.

The CV module (w18) allows you to integrate your Synth Kit with other analog synthesizers (for example, modular synths or analog keyboards).

The USB i/o module (w27) is an USB audio interface that will enable you to record directly into your DAW from the Synth Kit or send audio from your DAW to the Synth Kit for processing. Integrate your Synth Kit with common audio work stations (Ableton, ProTools, or other software like Traktor or Max/MSP).

Making Motion

There are a few output Bits that help you make your project more kinetic. So now that you know how to make music, you could make a project that dances to that music. First let's take a look at a few motors.

Vibration Motor

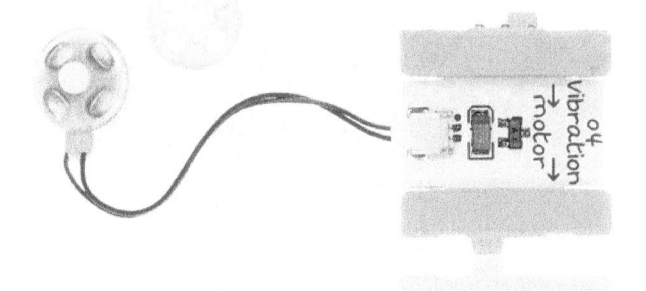

Inside your cell phone, there's probably a tiny vibration motor, just like the one attached to the vibration motor module (o4). This one will make things vibrate and buzz. The included vibeSnap connector helps you attach the vibration motor to the materials that you're working with. You can control the intensity of the vibration with any of the analog input Bits.

DC Motor

The DC motor Bit (o5) has a geared direct current motor attached to it. A switch sets the direction of the rotation on the motor's shaft and you can affect the speed of the motor by varying the voltage going into the Bit. Try that out now by connecting any analog input between a power module and the DC motor. The higher the voltage, the faster the motor goes.

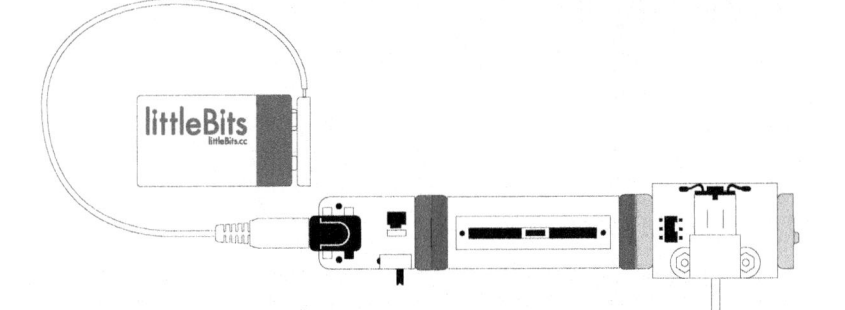

Figure 3-15. *The speed of the DC motor can be controlled by the amount of voltage going to it. You can vary that voltage with an analog input Bit.*

You may also notice that the shaft of the motor isn't exactly a circle, but it has a small flat side. This will help you attach all kinds of objects like cardboard, plastic wheels, clay, or food to the end of the motor.

The motor also comes with a motorMate, first introduced to you in "motorMate" on page 22. It'll give you more options for attaching various things to the DC motor. It's perfect for attaching a Lego axle to your DC motor.

Project Tutorial: GramoPaint by Makerspark

http://littlebits.cc/projects/gramopaint

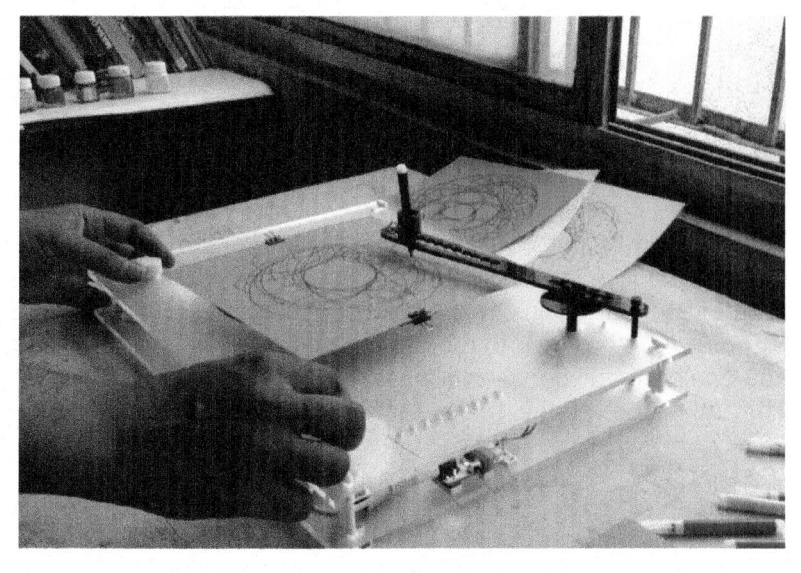

GramoPaint comes from a collective of makers in India. They used two DC motors, one to spin a plate which holds a piece of paper and another to move an arm that holds a marker. The resulting works have colorful spiraling patterns.

To help them execute the project, they used a laser cutter to create the flat pieces and a 3D printer for the spacers, mounts, knobs, and more. The full instructions and design files are available on the project page so that you can make your own.

Servo

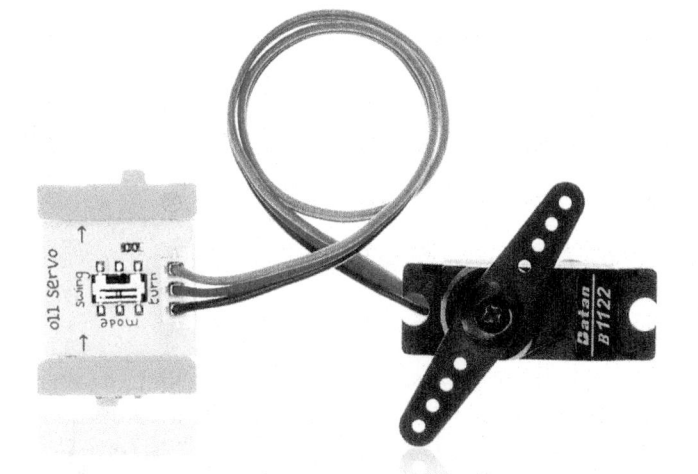

A DC motor and servo motor are very different from each other. Instead of spinning, the shaft of a servo motor simply holds at a particular angle. If you want to control a gripper claw, for instance, you could rig a servo motor to do that. The littleBits servo Bit (o11) has two modes controlled by a switch. In turn mode, the position of the shaft is determined by the voltage of the input signal. When the switch is set to swing mode, the servo swings back and forth. The speed will be determined by the input voltage.

Try connecting a dimmer or another analog input as shown in Figure 3-16 to try out both modes to get a sense of how they work.

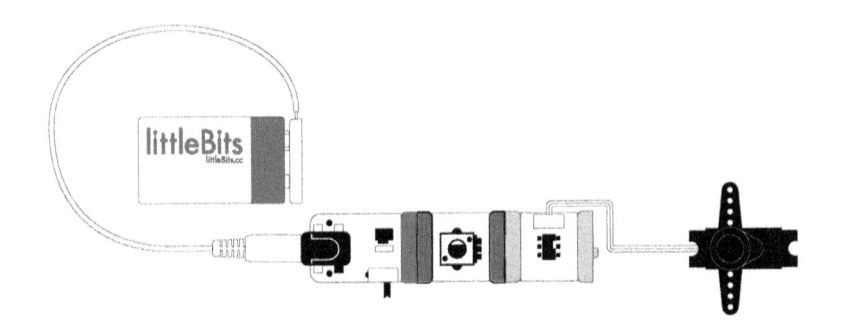

Figure 3-16. *The speed of the DC motor can be controlled by the amount of voltage going to it. You can vary that voltage with an analog input Bit.*

A servo *horn* is the piece that attaches to the shaft of the motor with a small screw. The servo Bit includes a few different horns, as shown in Figure 3-17. There are a lot of ways to have the servo attach to the moving parts in your project. For instance, you can connect a thin wire, like from a coat hanger into one of the holes on the horn.

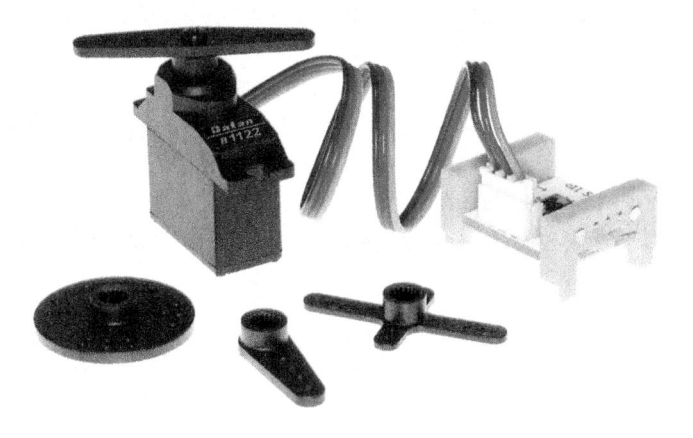

Figure 3-17. *The servo Bit includes attachments called horns.*

Project: RC Car

http://littlebits.cc/projects/remote-control-facetime-car

Using a pair of wireless modules, a couple DC motors, and a servo, you can create a remote-controlled car that can explore your world. Use the slide dimmer Bits on the controller to steer the vehicle and control its gripper arm.

Take it one step further by adding in two smartphones, and you can explore the unknown—like Mars or under the couch! Set up a videocall between two smartphones and place one on the vehicle and one in the controller. Drive your vehicle wherever you like and feel like you are riding along. Figure 3-18 shows the complete project.

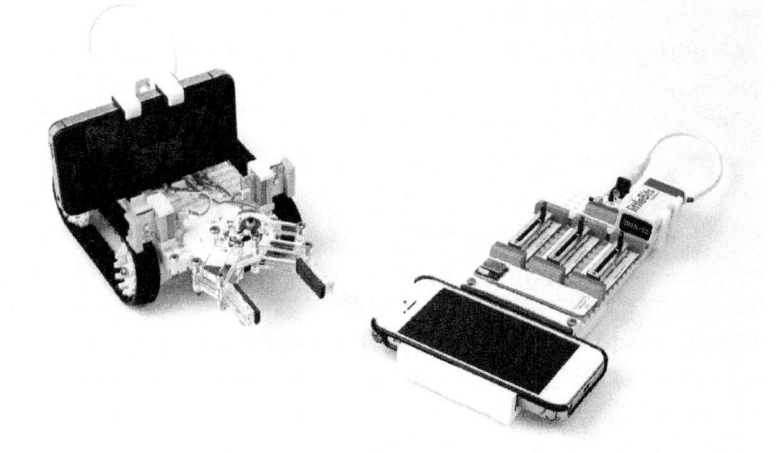

Figure 3-18. *The RC car and controller, with optional smart-phones*

To make this project, you'll need:

- 2 bright LED Bits
- 2 DC motor Bits
- 2 fork Bits
- 2 power Bits

- 2 9V batteries
- 1 RGB LED Bit
- 3 slide dimmer Bits
- 4 wire Bits
- 1 servo motor Bit
- 1 wireless receiver Bit
- 1 wireless transmitter Bit
- 2 mounting boards
- Assorted foam, acrylic board, wire, tape, etc. for the body
- Various screws
- RC car wheels
- Optional: 3D printer/laser cutter and supplies for both

If you don't have access to a 3D printer and laser cutter, you can fashion these parts out of foam, cardboard, and other materials. If you're not using a smartphone, you can improvise more with your gripper design.

Here's how to make the RC car. Start by building the circuits:

1. Build the controller circuit. Connect power, a fork Bit, and a slide dimmer on each branch (3 total) of the fork, along with a wireless transmitter, as shown in Figure 3-19.
2. Build the vehicle circuit. Connect power to a fork Bit. On the first and the third branch, place a wire and bright LED. On the second branch of the fork, add a wire and the wireless receiver, followed by a wire and DC motor on the first channel. The second wireless channel needs to have a servo (in turn mode), and on the third channel, connect an RGB LED and DC motor. Figure 3-20 shows the circuit.

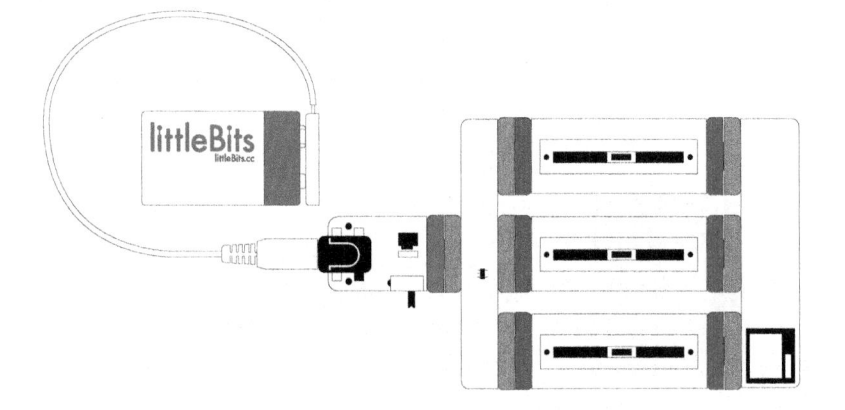

Figure 3-19. *RC controller circuit*

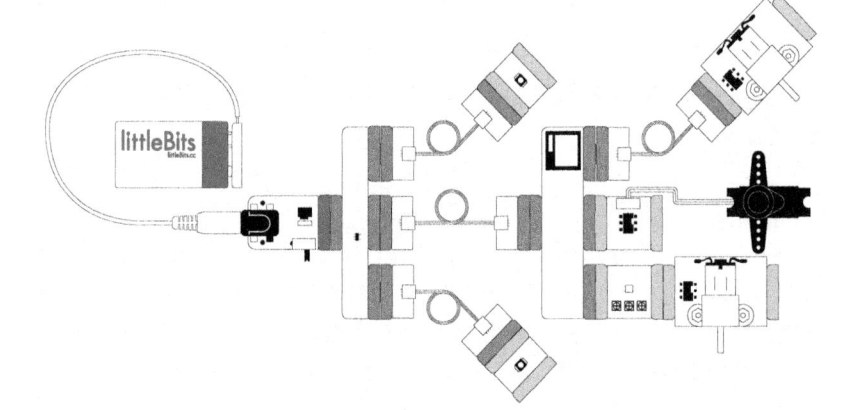

Figure 3-20. *RC car circuit*

On this project, you need to think about the user interface, since someone has to actually use it! The slide dimmers on the first and third channels control the two DC motors on corresponding receiver channel. It feels natural to have the right wheel's control on the right side and the left wheel control on the left side. It also makes more sense to position the sliders vertically so the car drives forward when you push the dimmers forward. See Figure 3-21 for a suggested layout.

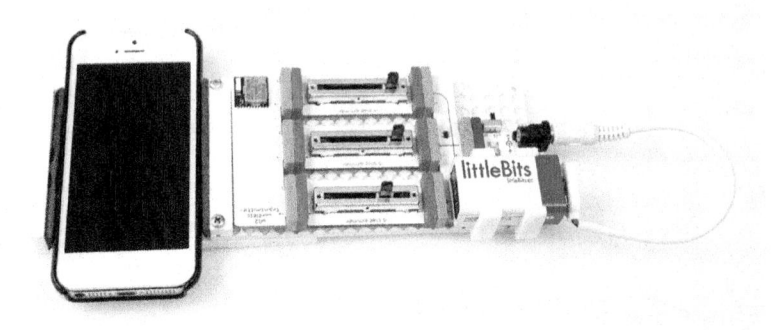

Figure 3-21. *The controller, shown with optional smartphone attachment*

Consider the layout of the circuit on your vehicle as well (especially the wireless receiver). Whether you put the DC motors at the front or back of your car or if you mount the modules on top of the car body or on the bottom. These all affect the way the vehicle moves. See what works best for you. Figure 3-22 shows one possible configuration.

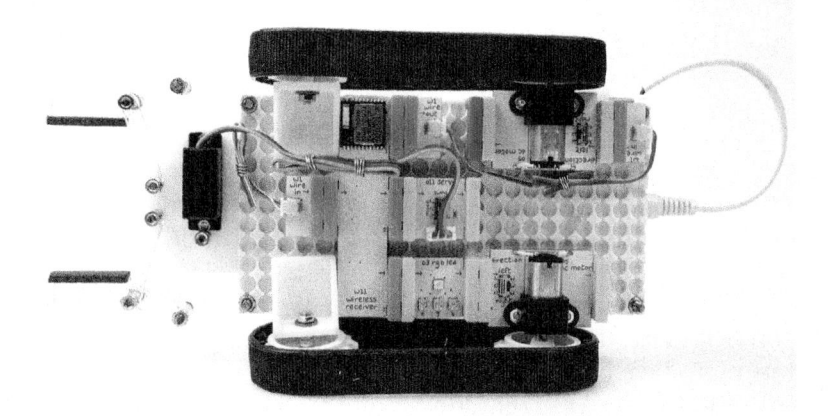

Figure 3-22. *Laying out the circuit underneath the car*

Now you need to make a gripper. Thingiverse user jjshortcut did a great job making a simple gripper using a micro servo (*http://www.thingiverse.com/thing:2415*). We modified this file a little to fit our servo, and also designed a mount that makes it possible to connect the gripper part to our mounting board directly. You can download these modified gripper parts and the gripper mount from Thingiverse (*http://www.thingiverse.com/thing:258971*). Once you laser cut or 3d-print these parts, assemble them all together. Put the left arm of the gripper directly on the servo shaft. Use two M2 screws with counter-sink to hold the servo motor. All the other screws should be M3. Before you slip the gripper into a mounting board, we suggest you use some thick gluedots or VHB tape on the flat face between them.

Build the body and assemble the tracks. We used the Pololu 30T track set (*http://www.pololu.com/product/1416*):

1. Sandwich two mounting boards together with the module side facing out.

2. Screw them together with M3 mechanical screws and nuts.

3. Plug the drive sprocket from the Pololu track set into DC motor shaft.

4. Fix the idler sprocket onto a 3d-printed axle-hole part—you will need two of these (*http://www.thingiverse.com/thing: 258997*).

5. Place the tracks with DC motor modules and 3d-printed parts on the bottom of the two mounting boards.

After some trial and error, we found that the track performs great when the distance between DC motor module and 3d-printed axle hole part equals 7 mounting holes (about 12.5 mounting holes from center to center). We recommend you use some 3M VHB tape or thick gluedots in between these parts and the mounting board because tension from the rubbery track may pull the modules off. Don't forget to check that your vehicle moves properly in accordance to your controller. We placed the drive wheels at the rear and set the right DC motor to left mode. The left DC motor is set to right mode.

Place all the circuits on the vehicle. Print out 2 vertical bit mounts (*http://www.thingiverse.com/thing:243540*) to stand bright LEDs upright and print 2 battery holders (*http://www.thingiverse.com/thing:243530*) to hold batteries in place. Be wise and creative using given space on two mounting boards.

If you're planning to use a pair of smartphones with this project, you have a little bit more to do:

1. Print out the smartphone holder (*http://www.thingi-verse.com/thing:259039*) and put it on the vehicle's back. Use three M3 screws to make it adjustable to hold any phone.

2. Finish the controller. With the controller circuit on a mounting board, you're ready to add the smartphone holder. Using a laser cutter and the file provided, make a smartphone holder for the controller. The big plate with slits on the sides goes under the mounting board, and the small piece with screw holes goes over the mounting board. Screw them together when your phone is in place. You can unscrew and slide the plate to adjust the size according to the size of your phone.

3. Put some pieces of EVA foam on the tips of the gripper arm, smartphone holder on the vehicle, and the smartphone

holder on the remote controller. This may prevent your precious phone from slipping out and falling to the ground.

4. Ask a friend to borrow a smartphone and then call yourself though a videocall (like Facetime or Skype). Place one phone on the vehicle and the other on the remote control.

You are now ready for your great expedition! Keep in mind that you can't drive this car in reverse, so give yourself plenty of turning radius or you'll be moving your couch to retrieve a lost car!

4/Wireless and Cloud Communication

So far, the Bits you've been working with need to physically connect to each other in order to pass the signal from one Bit to the next. But with the help of the wireless receiver and transmitter, you can send the signal without wires. And with the cloudBit, you can connect your project to a WiFi network so that it can be accessed wirelessly via the Internet. If you have a pair of cloudBits, you can have one pass its signal through the Internet. This means your project can stretch across the globe.

Wireless Transmitter and Receiver

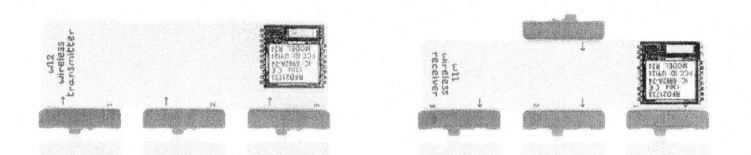

The wireless transmitter (w12) and receiver (w11) Bits work with each other in order to easily make your project work without need for the bits to be right next to each other. The transmitter communicates the incoming signals directly to the receiver, so it works without the need for a wireless network and without any kind of setup. These Bits are great because they're plug and play.

With the wireless Bits, you can make a remote control for a robot or car, create a wireless door lock, or be alerted in your living room when your back door is opened.

To try out the transmitter and receiver:

1. Connect a power Bit to a button and connect the button to channel 1 of the transmitter, as shown in Figure 4-1.

2. Connect a second power Bit to a receiver and connect an LED (or any other output) to channel 1 of the receiver, as shown in Figure 4-1.

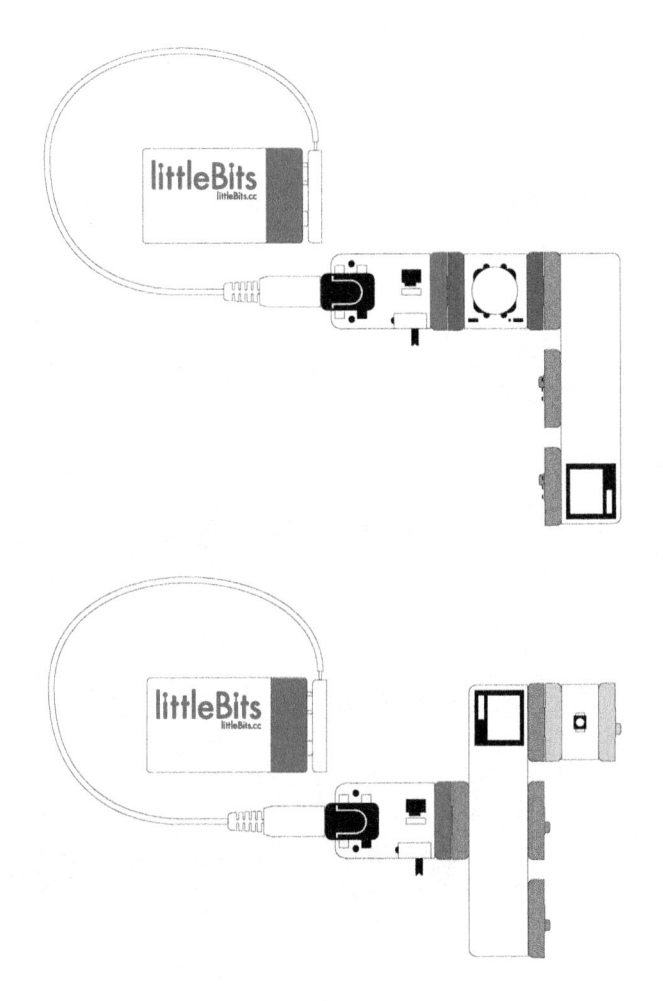

Figure 4-1. *Trying out the wireless transmitter and receiver modules with a basic button and LED.*

When you push the button on one circuit, the output should activate on the other. Try seeing how far you can go and still trigger the output with the button. The range of the wireless Bits is about 100 feet, so that it should still work from the next room, but your mileage may vary since there are a lot of factors that affect its performance.

With three channels, you can have three different inputs trigger three different outputs with just a single transmitter and receiver pair. Try adding other inputs to channel 2 and 3 of the transmitter and outputs to channel 2 and 3 of the receiver to test this out (Figure 4-2).

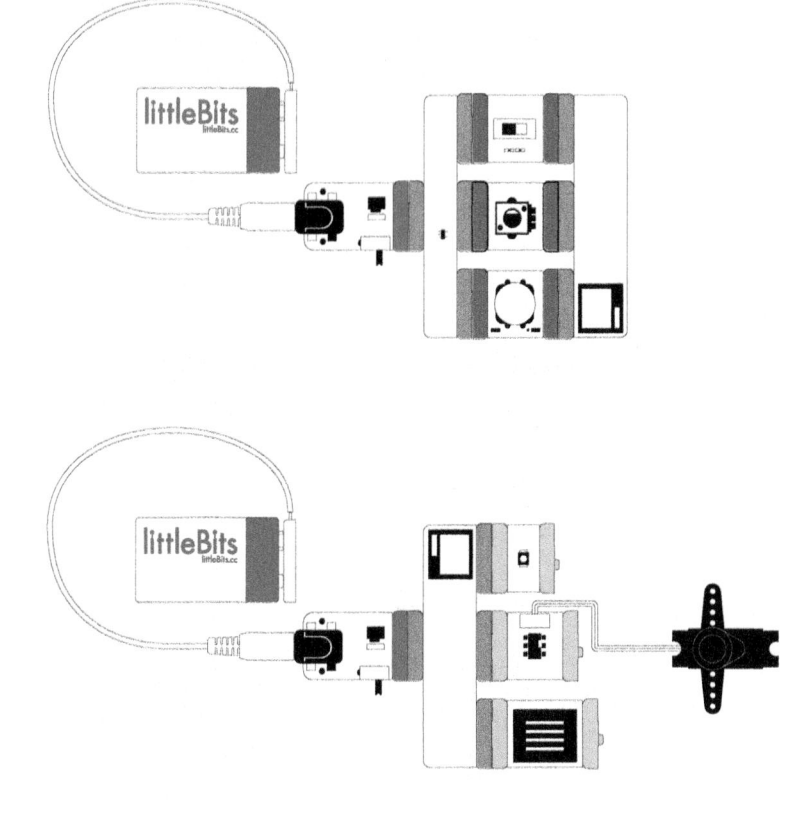

Figure 4-2. *Three inputs on the transmitter can wirelessly control three outputs on the receiver.*

A transmitter Bit will wirelessly broadcast the signal from the three inputs to any receiver. That means that if you have one transmitter and multiple receivers, all of the receivers will get their marching orders from that transmitter. However, this also means that you can only use one wireless transmitter within its range, otherwise you'll encounter some unexpected behavior.

Project Tutorial: Remote Control Facetime Car

http://littlebits.cc/projects/remote-control-facetime-car

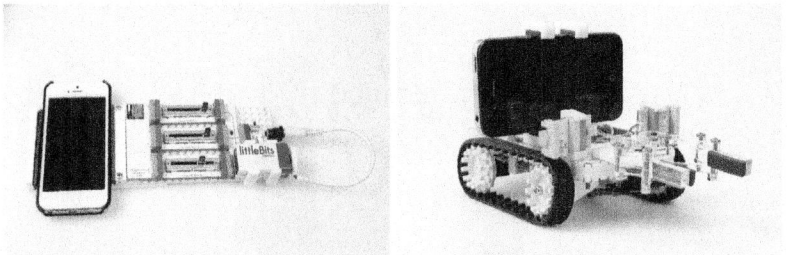

Using a pair of wireless modules, a couple DC motors, a servo, and two smartphones, you can explore the unknown—like Mars or the universe under a couch! Here is a solution for a future astronaut or a person desperate to find lost quarters. Set up a video call between two smartphones and place one on the vehicle and one in the controller. Use the slide dimmers on the controller to steer the vehicle and control the gripper arm. Drive your vehicle wherever you like and feel like you are riding along. See "Project: RC Car" on page 87 for all the details on making this yourself.

Remote Trigger

There's another simple way to control your littleBits projects from across the room with the help of the remote trigger (i17). The remote trigger lets you use a common remote control with your modules. Make your littleBits circuit and point your remote control at the remote trigger's sensor. Then, press any button on your remote control to activate the modules after the remote trigger. The remote trigger will work with almost any button on a remote that uses infrared light to send signals. You can also use the IR LED (o8) to activate the remote trigger.

IR Transmitter and AC Switch

The AC switch and IR transmitter can be used to remotely toggle on or off any device that plugs into a standard household outlet. You can retrofit your household appliances to make them "smart" and allow you to control them remotely. Both the AC switch and IR transmitter are included in the Smart Home Kit (*http://littlebits.cc/kits/smart-home-kit*).

The AC switch can power devices which use up to 15A, such as lamps, fans, coffee makers and air conditioners.

The IR transmitter sends a short pulse of modulated infrared light. Use it to wirelessly activate the AC switch to turn appliances like a lamp or fan on and off. You can activate it with either a momentary (button) or latching (toggle switch) input. To pair, press the button on the AC switch until the red LED blinks. When you select a channel on the IR Transmitter and activate it with a button or other input, the AC switch will remember that channel until it is reprogrammed the same way.

You can use them with a sound trigger (i20) to make your own clapper machine (*http://littlebits.cc/projects/clapper-machine*) and turn on any household object by clapping your hands. For more ideas, check out the Tips & Tricks (*http://littlebits.cc/tips-tricks/tips-tricks-ir-transmitter-the-ac-switch-2*).

cloudBit

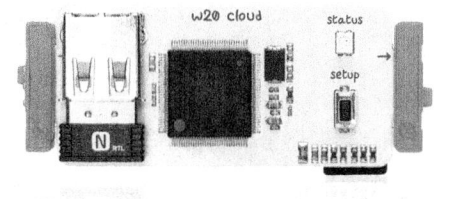

When the things you make can connect to the Internet, you unlock an enormous amount of power. With the cloudBit (w20), you get all that power in a very easy-to-use Bit. Not only does the cloudBit enable long distance communication between littleBits projects, but it can also connect to a multitude of online services through IFTTT (more on this in "IFTTT" on page 112). And if you're an advanced developer, its open API means that the possibilities with how you can use it are limitless.

The cloudBit works by connecting to the Internet over your WiFi network. Through the Internet, it establishes a connection to the littleBits Cloud Control servers, which receive signals from the cloudBit's input bitSnap connector and send signals to the output bitSnap connector. The littleBits Cloud Control server will help you connect your cloudBit to other cloudBits, to IFTTT, and to your own web servers with its API.

Getting Set Up

In order to use your cloudBit, you will need a littleBits account and a WiFi connection. In the setup process you will give your cloudBit a name and tell it which WiFi network to connect to. It will then remember that network and connect to it automatically whenever it's powered on. This process is really straightforward, and the littleBits cloud control application walks you through it. For the sake of completeness, here's how to program your cloudBit with your WiFi network information.

1. Connect your cloudBit to a USB power Bit. The cloudBit will first show a white light as it boots.

 You must use the USB power Bit to power the cloudBit. The other power Bits may not provide enough electrical current and your cloudBit will act erratically. The USB power Bit must also be connected to a power source that can provide at least 1 amp.

2. While it's booting, open your computer, phone, or tablet's web browser and go to the address *http://control.littlebitscloud.cc*.

 Your device must be on the same WiFi network that you plan to connect your cloudBit to.

3. If you don't already have a littleBits account, you'll need to create one. Otherwise log into your account. If you're already logged in to your littleBits account, you won't need to log in again.

4. Give your cloudBit a name and click Save Name:

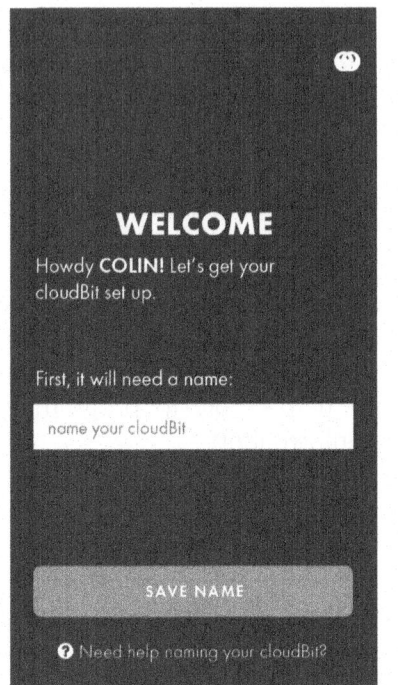

5. Wait for the status light to start blinking, and click "STATUS LIGHT IS FLASHING."

 These instructions are for a brand new cloudBit that hasn't been associated with a WiFi network yet. If your cloudBit has been associated with a littleBits account and WiFi network already, the process will be slightly different.

---

6. Put your cloudBit in setup mode. Hold down the setup button on the cloudBit until it blinks blue, then let go. When the blue light turns steady, click the button that says "BLUE LIGHT IS STEADY." When the blue light is steady, it means that the cloudBit is acting like a WiFi access point.

7. Go to your computer or device's WiFi settings to connect to this network. It will be named "littleBits_Cloud_..." (The last

few characters in the access point name will be a unique string of letters and numbers for your cloudBit.):

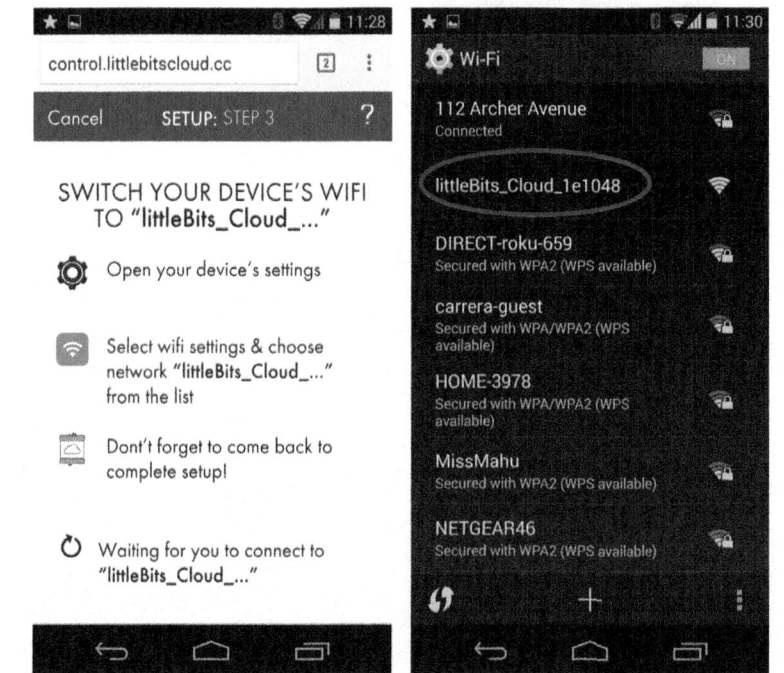

8. Go back to your browser and select your home WiFi from the list and enter your password. It may take a minute or two for this step, so be patient.

✓ ─── ✓ ─── ✓ ─── ④ ⋯ ⑤

PAIR YOUR CLOUDBIT

Tell your cloudBit what local wifi to
connect to by selecting it below:

CHOOSE A NETWORK...

random network 1   🔒 🛜

My Wifi Network   🔒 🛜

random network 3   🛜

random network 3   🔒 🛜

random network 3   🔒 🛜

random network 4   🔒 🛜

9. Connect back to your regular WiFi connection in your computer or device's settings.

10. At first the cloudBit will blink green, but it will turn steady green as soon as it's connected to the WiFi network and can access the Internet.

11. Once your cloudBit connects, you'll be automatically moved on to test it.

AWESOME!

You have successfully connected your
cloudBit to the internet.

New to the cloudBit?

TRY SOME SAMPLE CIRCUITS

Returning expert?

GO TO DASHBOARD

If there is a connection problem, you'll be able to go back to connect your cloudBit in "Diagnostic Mode" to get specific troubleshooting information.

To test the cloudBit:

1. As instructed, connect a button between the power Bit and the cloudBit and wait for the light to turn green again. Since you're removing power from the cloudBit and reconnecting it, it needs to warm up and associate with your WiFi network (blinking green indicates it's connecting to the WiFi; if the light ever turns red, please contact littleBits support). Go to the next screen by dragging or swiping to the left.

2. When you press and hold the button, you should see the needle go from zero to 100%. Go to the next screen by dragging or swiping to the left.

3. As instructed, add an output Bit to the output of the cloudBit. Go to the next screen by dragging or swiping to the left.

4. Press the button on your device to see the output Bit activate!

After the setup process, you'll be brought to your Cloud Control dashboard.

Cloud Control

Cloud Control is your Web-based dashboard for all your cloudBits. It works great on a computer, tablet, or phone's Web browser and is accessed by going to *http://control.littlebitscloud.cc*. Best of all, no matter where you are in the world, as long as you have an Internet connection, you can access your cloudBit to read the state of its input or to control the output.

Figure 4-3. *From Cloud Control, you can view all your cloudBits, adjust their settings, send signals to them, and view their input signals.*

Any cloudBit that you set up with your littleBits account will appear under "My cloudBits" on the left-hand side. Buttons along the bottom of the app let you control the output, read the input, connect to IFTTT (covered in "IFTTT" on page 112), and change the settings of your Bit.

Take a moment to explore Cloud Control:

1. Make sure that the cloudBit you just set up is selected on the left-hand side, under "MY CLOUDBITS."

2. Click or tap the SEND button on the bottom.

3. The first send screen is a big purple button. Whenever you push it, it will turn on the output of the cloudBit for one second:

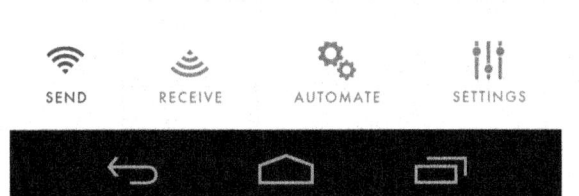

4. Swipe or drag to the left and you'll see a slider control with the number 0. This is how you can send an analog signal out of the cloudBit's output. Drag the slider and watch the LED you attached change in brightness!

5. Click or tap the RECEIVE button on the bottom. Here there are two visualisations for the input of the cloudBit, both showing the full range of analog values it can receive. The first is a needle gauge. A swipe or drag to the left will reveal a numerical readout.

6. With the button connected, you'll notice that it's either entirely off, or entirely on. Try connecting an analog input before the cloudBit and letting it boot up again. You'll see how sending an analog signal to the cloudBit has an effect on the readouts in Cloud Control.

7. Click or tap the SETTINGS button on the bottom of the screen. Here you can change the name of your cloudBit, change its WiFi network, view advanced information about it, delete it from Cloud Control, and set it to follow another Bit (see "Project: Thinking Of You" on page 115).

8. The AUTOMATE button on the bottom of the screen takes you to the littleBits channel on IFTTT. There's a lot to explore there, so let's dive into that in the next section.

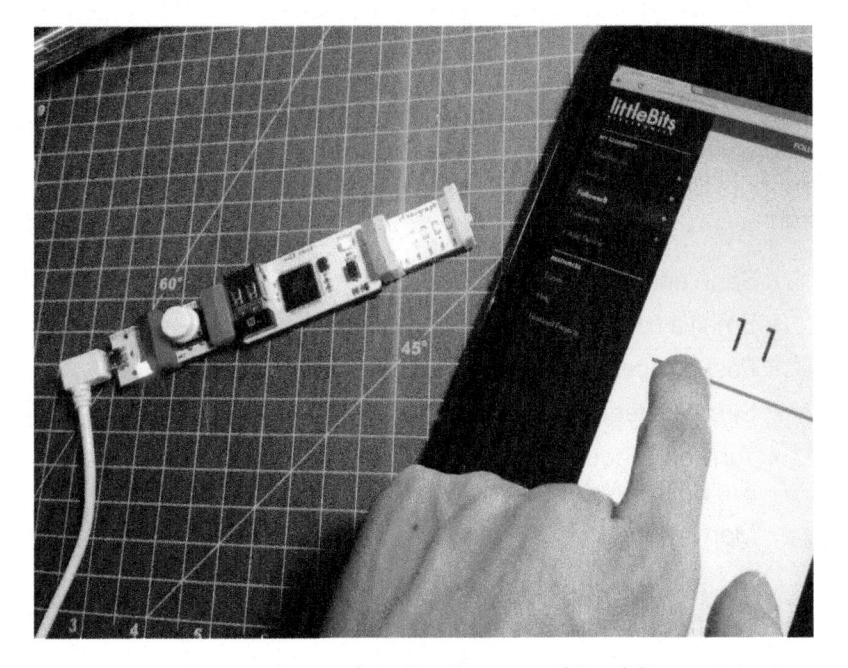

Figure 4-4. *Testing out a cloudBit from the Cloud Control dashboard.*

IFTTT

IFTTT (pronounced like "gift" without the "g.") stands for "If This Then That". It's a free service that helps you to connect other online services together. For instance, you can set up IFTTT so that if your favorite blog posts a new update, then it will send you an email. Or if you like a photo on Instagram, then automatically tweet it to your followers.

IFTTT not only connects online services together, but also works with physical products that have a network connection. For example, there's an app for iPhone and Android to receive notifications or to trigger actions on other services. If you have any of their supported activity trackers, home automation, or other connected devices, they can trigger or be triggered by other services through IFTTT.

The littleBits cloudBit works like a champion with IFTTT. Inputs into the cloudBit can be used to trigger a multitude of different actions on many online services. If you connect a simple button module to the cloudBit and used IFTTT, you could have that button:

- Send an alert to your phone
- Email a friend
- Post a Facebook status message
- Send a tweet
- Turn on a house light through home automation platforms like Belkin WeMo, Quirky Wink, or Philips Hue
- Many more

Actions from other online services can trigger the cloudBit's output. If you connect a simple LED to the cloudBit and used IFTTT, you could have the following things turn the LED on:

- You receive an email
- The forecast predicts rain for today
- Your favorite team is playing
- The International Space Station is passing over you

- Your best friend posts a photo to Instagram
- Many more

To try out IFTTT with the cloudBit, you can configure it to have a simple button press send an email to you:

1. If you don't have an account already, create one at *https://ifttt.com/*
2. Once in your account, click "Channels" and then choose the littleBits channel.
3. When you're viewing the channel, click "Activate."
4. Log into your littleBits account if it asks you to.
5. Click "My Recipes" at the top.
6. Click "Create a Recipe." A recipe is a single trigger and action paired together.
7. It first asks for a trigger channel. Choose the littleBits channel.
8. Select the "input received" trigger.
9. Select the cloudBit name you chose in "Getting Set Up" on page 103 and click "Create Trigger."
10. Now it asks for the action channel. Choose email.
11. Click the "Send me an email" action.
12. If you want, customize the text of the email.
13. Click "Create Action."
14. Press the button connected to your cloudBit.

Quite soon after you press the button, you should see a new mail from IFTTT in your inbox. It might not seem amazing at first, but doing exactly this would be very tough without the power of littleBits and IFTTT.

When you combine the modular power of littleBits and IFTTT together, there's so much you can do! Here are a few projects from the littleBits community that take advantage of that power:

Project: The Game is On!

If you're an absent-minded sports fan, you might forget when your team is playing. Using the ESPN channel on IFTTT, your sports memorabilia will light up when your team takes the field.

To make this project, you'll need:

- cloudBit
- USB power Bit
- USB power adapter
- micro USB cable
- light wire Bit
- A piece of team memorabilia or a printout of your team's mascot or logo
- Cardboard or a picture frame, if needed.

Here's how to set this project up:

1. Follow the instructions in "Getting Set Up" on page 103 to set up your cloudBit to connect to your WiFi network.
2. Follow the instructions in "IFTTT" on page 112 to associate your account with IFTTT.
3. Log into your IFTTT account and click Create Recipe.
4. Choose ESPN as the trigger channel.
5. Choose the trigger "New game start" and then select your league and team.
6. Click Create Trigger.
7. Choose littleBits as the action channel.
8. Choose the cloudBit to use for the project, set the output level to 100% and duration to Forever. Click Create Action.
9. Click Create Recipe.
10. Create another recipe: set the trigger to ESPN's "New final score" and the action to set the output to 0%.
11. Decorate your team memorabilia with the light wire and connect the Bit to the output of the cloudBit.

Now whenever your team starts a game, your memorabilia will light up. When the game's over, it turns off! If you want something less subtle, you can set the new game start trigger to turn on a buzzer for ten seconds. That way, you'll be sure you won't miss it!

Project: Thinking Of You

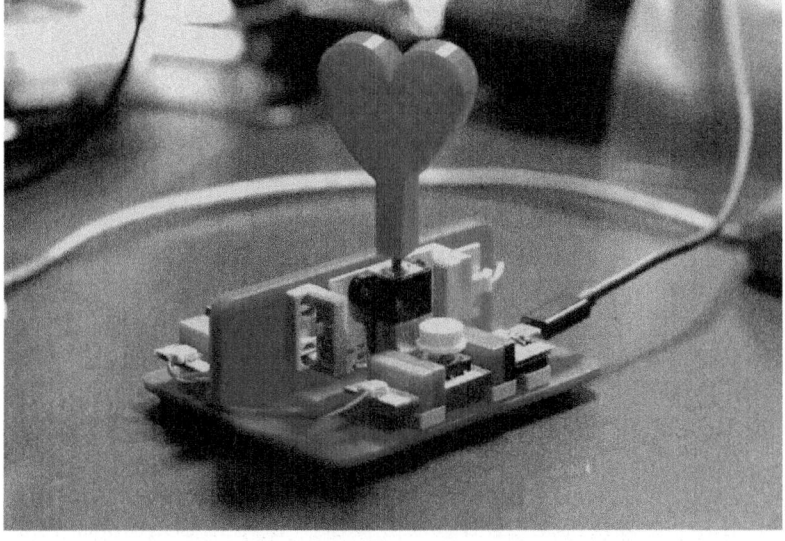

When you're away from someone you love, how can you show them that you're thinking of them? You can send a text message, but what about something even quicker? This project uses two cloudBits that communicate with each other so that your boo can know you're thinking of them and vice versa.

There are two parts to the project. One sits on your desk and the other sits on your loved one's, wherever in the world it is. When you push the button on your desk, the heart will spin on your partner's. When he or she pushes the button on their desk, the heart will spin on yours. It's an easy way to say "I'm thinking of you!"

This project is possible because of the cloudBit's *follow* feature, which is really neat. When you set one cloudBit (let's say cloud-Bit_A) to follow another (cloudBit_B) the output of cloudBit_A will match the input of the cloudBit_B. cloudBits can follow each other so that the input of cloudBit_A will be outputted on cloud-Bit_B. And the input of cloudBit_B will be outputted on cloud-Bit_A. It's like having a really long wire Bit that can stretch across the globe!

To make this project, you'll need:

- 2 cloudBits
- 2 USB power Bits
- 2 USB power adapters
- 2 micro USB Cables
- 2 button Bits
- 2 DC Motor Bits
- Clay, cardboard, or a 3D printer+filament for the heart
- Materials for the enclosure: Legos, cardboard, wood, or plastic will all work.

Here's how to make Thinking of You:

1. Connect the USB power, button, cloudBit and DC motors as shown in Figure 4-5.

2. Fashion a small heart out of clay, cardboard, or 3D print one that fits onto the DC motor's shaft. You can download a 3D-printable design file that attaches to the DC motor Bit from *http://www.thingiverse.com/thing:497720*.

3. In each location, follow the instructions in "Getting Set Up" on page 103 to setup the cloudBit to connect to the local WiFi network. Name each cloudBit something meaningful like your names.

Both cloudBits must be associated with the same littleBits account in order to use the follow feature.

----

4. Log into Cloud Control. For each cloudBit, click Settings and set it to follow the other cloudBit.

5. Use Legos, cardboard, wood, or any material you'd like to create an enclosure for each of the devices.

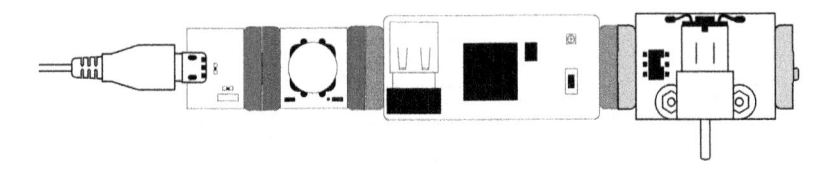

Figure 4-5. *Hooking up the cloudBits and motors*

As long as each device has a wifi connection, press and hold the button one device and the heart on the other one will spin.

Going Further with the cloudBit

If you're familiar with web development, you can take the cloudBit's power even further with the BitCloud API (*http://developer.littlebitscloud.cc/*). The API will let you access the full functionality, enabling you to read and write signals to the cloudBit from your own web server. You can even provide a callback URL and subscribe to events such as changes in state.

If you use Node.js to develop for the Web, the examples and resources from the littleBits workshop at NodeConf EU 2014 (*https://github.com/littlebits/cloud-api-lessons*) will help you understand how to connect to your cloudBits.

5/Programming with the Arduino Bit

As you've seen so far, there's a lot you can do with littleBits without writing a single line of code. But when you add the Arduino module to your projects, it unlocks the power of programmability. With a few lines of code, you can have your project behaving exactly the way you want.

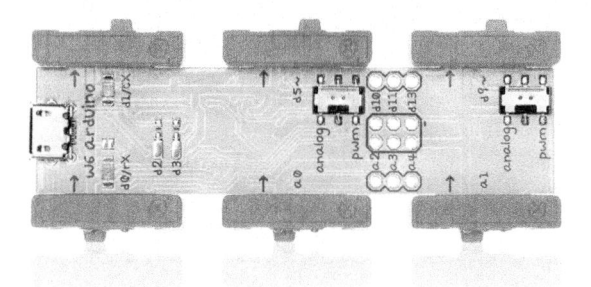

The Arduino module has three input and three output bitSnap connectors. On the top, there's a microUSB jack for programming the Bit from your computer. For the advanced makers, there are points where you can solder on connectors to get access to more inputs and outputs.

This chapter will get you started with the Arduino module, but there's no way we can cover everything that the Arduino is capable of. Luckily, there are plenty of amazing resources for helping you to do what you want. The official Arduino site (*http://www.arduino.cc/*) is a great place to start exploring the potential of the platform.

What is Arduino?

Arduino is a *programmable microcontroller platform*. At the core of the platform is a chip that you can program to read inputs, make decisions, and control outputs determined by the code that you write for it. The Arduino Uno you see in Figure 5-1 is an example of a basic Arduino board (and there are plenty more varieties). You can connect it to your computer to power it, upload code, and get data from the board. Arduino is also the name of the programming language and development environment that people use to program the board.

Figure 5-1. *The Arduino Uno is just one type of Arduino board, but the most common. The littleBits Arduino at Heart module is most similar to the Arduino Leonardo.*

Atmel's ATmega 32u4 microprocessor chip is the brains of the littleBits Arduino module, just like with the Arduino Leonardo and Micro. If you're already familiar with the Arduino Uno's behavior, there are a few key differences with working with this chip. The folks at Arduino have outlined them here: *http://arduino.cc/en/Guide/ArduinoLeonardoMicro*

The littleBits Arduino Module is an *Arduino at Heart* product, which means that even though it doesn't look exactly like an Arduino, it's compatible in many ways with the Arduino ecosystem. For instance, you'll use the standard Arduino code and Arduino software to program the Bit.

The littleBits Arduino module takes all the power and flexibility of Arduino and adapts it for the littleBits library, so that you don't have to worry about wiring or soldering. Just snap the modules you want to work with onto the Arduino module, program the Bit, and you're pretty much good to go.

Getting Set Up

In order to program the chip on the Arduino, you'll need to download the Arduino integrated development environment, or IDE. It's the tool you'll use to write your code, compile it, and upload it to the board. Follow the steps below to set up your computer to program the Arduino module and to test it out:

 A freshly unboxed Arduino at Heart module has the following behaviors, depending on which input you power the module through:

- If a signal goes to d0, the Arduino module creates a pulse every second to any output Bit you connect to d1.
- Any input you send to a0 is read and directly output to d5.
- If an analog input signal goes to a1, its setting determines the fade rate through d9.

As soon as you reprogram the Arduino at Heart module, these behaviors are replaced with the code you load on it.

1. Connect a power module to one of the input bitSnap connectors on the Arduino Bit.

The Arduino module gets its power from the bitSnap connectors, not from USB, as shown in Figure 5-2 . A power module will need to be connected to the Arduino module when you're programming it.

2. Connect any output Bit (such as an LED) to the bitSnap connector labeled **d1/tx**. If you're using a fresh Arduino Bit straight from the littleBits factory, you should see the LED toggling on and off once per second. This is because the chip on the Bit is initially programmed to do this.

3. Connect any output Bit that can show analog output (such as bargraph module) to the output bitSnap connector labeled **d9~**. Again, since the Arduino Bit is programmed before it leaves the factory, you should see the bargraph slowly fill up and then drop back down repeatedly.

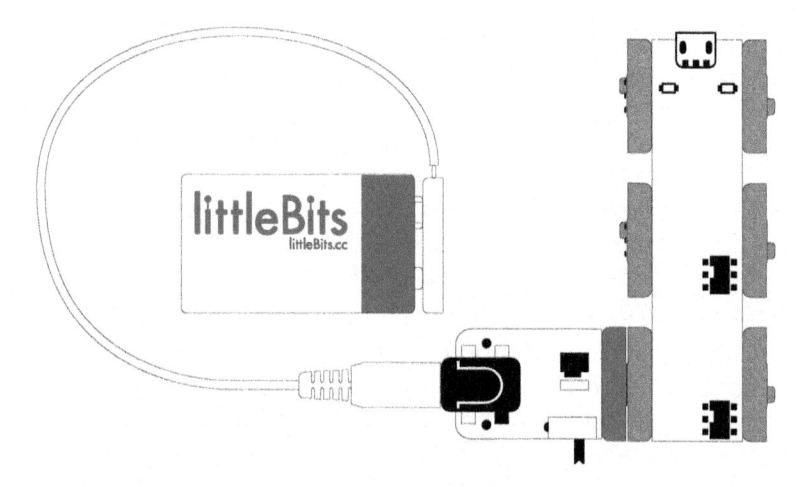

Figure 5-2. *Powering the Arduino module*

Next, install the software and program the board:

1. Go to the Arduino downloads page at *http://arduino.cc/en/Main/Software*.

2. Click the link to download the latest version to the installer for your operating system. There are versions of the IDE for Windows, Linux, and OS X.

3. Execute the installer or copy the application to your computer's drive (this step will vary depending on your operating system).

4. Launch the Arduino IDE (this step will vary depending on your operating system).

5. Using a microUSB cable, connect the Arduino bit to your computer. Your setup should look something like Figure 5-3. If you're on Windows, you may need to follow the Windows driver installation instructions (*http://arduino.cc/en/Guide/ ArduinoLeonardoMicro?from=Guide.ArduinoLeo nardo#toc10*).

6. Click File→New to create a new sketch. A sketch is another word for "program" in the world of Arduino.

7. Enter the code from Example 5-1 into the code window. It should look like Figure 5-4.

8. Go to Tools→Board and select "Arduino Leonardo."

9. Go to Tools→Serial Port and select the port that the Arduino Bit is connected to. It will be something like COM3 on Windows, dev/ttyUSB... or /dev/ttyACM... on Linux, and /dev/ tty.usbmodem... on Mac OS. The exact name will depend on the particulars of your computer. You may need to do some trial and error to find the right serial port (unplug the module from the computer, check the list of ports, plug it in, and select the one that wasn't there when it was unplugged).

10. Click the Upload button on the toolbar. You will see a little bar on the bottom of the Arduino IDE that shows the progress of the upload. The upload should be pretty quick (less than 10 seconds or so). Once uploaded, the progress bar should say "done uploading."

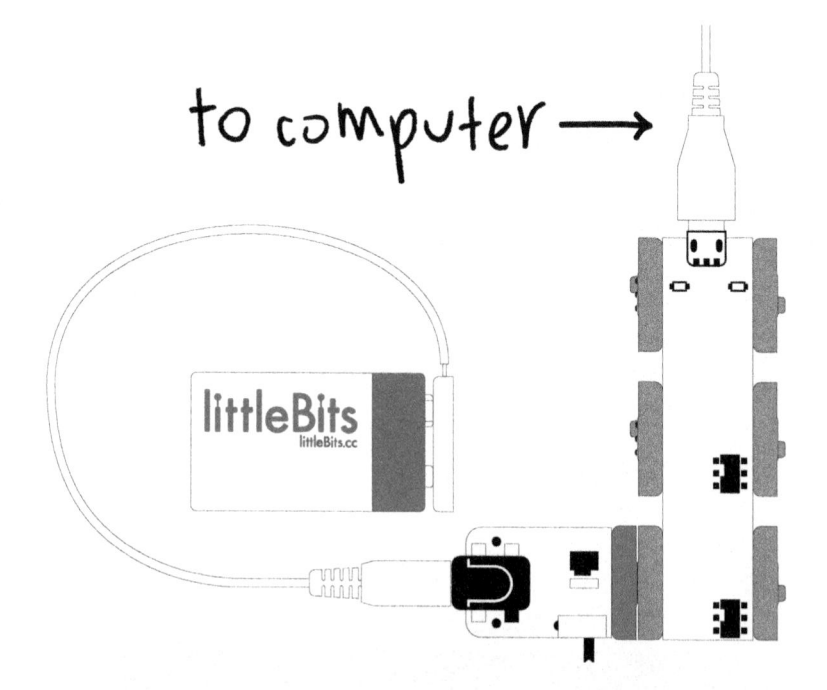

Figure 5-3. *The Arduino module connected to a computer*

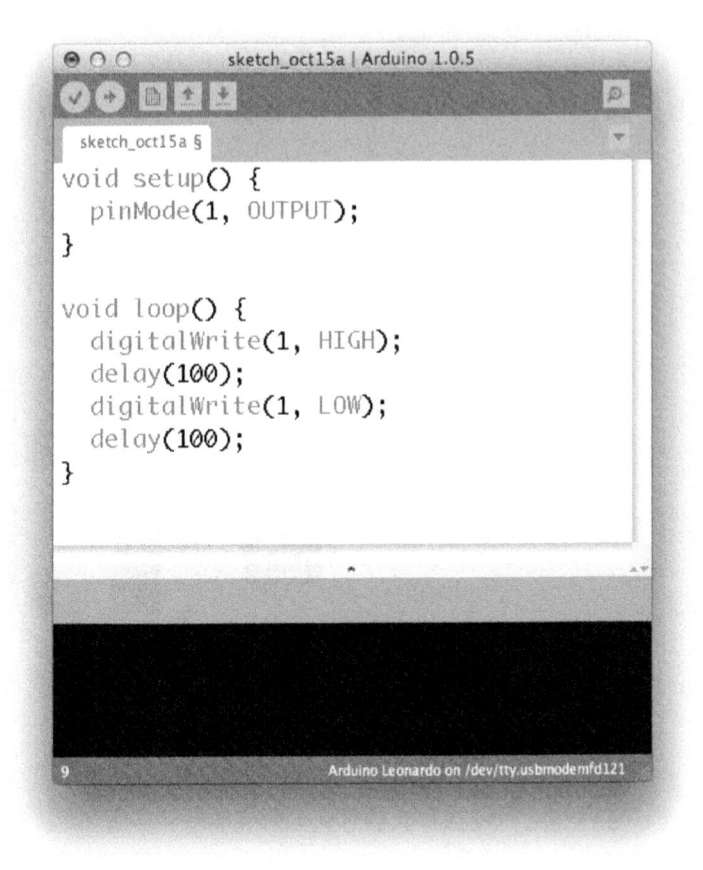

Figure 5-4. *The Arduino IDE*

If you need more guidance on using Arduino on your platform, refer to the Arduino Leonardo and Micro, see the official guide for the Arduino Leonardo and Micro (*http://arduino.cc/en/Guide/ArduinoLeonardo Micro*).

If you see the LED connected to d1/tx blinking very fast, you'll know that you've done everything right and you're all set up to program the Arduino. This blinking represents the serial communication between your computer and the Arduino module. Any time you upload new code to the Arduino module, it replaces the code that's already on it. The code that you upload will be held in the board's memory and the code will run automatically any time you supply power to the board.

Example 5-1. Your first Arduino program, Blink

```
void setup() {  ❶
  pinMode(1, OUTPUT);  ❷
}

void loop() {  ❸
  digitalWrite(1, HIGH);  ❹
  delay(100);  ❺
  digitalWrite(1, LOW);  ❻
  delay(100);  ❼
}
```

❶ All the code in the setup function is run once when the module is first powered on or reset.

❷ Set up pin d1 to be an output.

❸ All the code in the loop function is run repeatedly after the setup function completes.

❹ Turn pin d1 ON.

❺ Wait for one tenth of a second (100 milliseconds).

❻ Turn pin d1 OFF.

❼ Wait for one tenth of a second (100 milliseconds).

Arduino Sketch Basics

Let's take a closer look at the code that makes up an Arduino sketch.

Every Arduino sketch has two main blocks of code: **setup** function and a **loop** function. The code inside the setup function is executed once when the Arduino Bit is powered on. The code inside the loop function is executed repeatedly until power is disconnected

The Arduino code you write is *procedural*, which means that within each of these functions, each line of code will be executed one at a time in the order it is written. Your code is essentially a list of steps to be carried out by the microcontroller chip on the Arduino Bit.

Usually, each statement, or instruction, is on its own line, but this is simply meant to help us mere humans read the code. You don't actually need to do this. However, you do need to end each statement with a semicolon (;). When you're programming an Arduino, it's like the period at the end of a sentence.

It's also helpful to know how to insert comments into your code. Comments (Example 5-2) help other people who look at your code to understand what you were intending to do. You also help yourself by adding comments so that you can go back to code that you wrote previously and figure out what you were thinking!

Example 5-2. Two kinds of comments

```
pinMode(d1, OUTPUT); // anything after the slashes on a line
                     // is ignored.
/*
Anything between a slash asterisk and
an asterisk slash will be ignored by the compiler.
It's to use for great for multi-line comments.
*/
```

You learned about the special **setup** and **loop** functions. These are blocks of code that you'll write the definition of when you create your own sketches. As a part of the Arduino code library, there are also many functions that have already been written that you can simply execute from within your code. Let's take a closer look at three of them now:

pinMode

pinMode is a function that's usually called within your **setup** function. It tells the microprocessor how you intend to use a particular pin, as an input or an output. It's important to know that the microprocessor on board has many pins that can act in different ways so you have to tell it how you intend to use a pin (in this case, the pin is the bitSnap connector). The syntax for `pinMode` is:

```
pinMode(pin, INPUT or OUTPUT);
```

In the parentheses after the word `pinMode` (case sensitive!), there are two parameters for this function, separated by a comma. The first parameter is the pin and the second parameter sets it as either input or output. For pins d1, d5, or d9, it'll usually be OUTPUT, as those are the three output bitSnap connectors on the module. For pins d0, a0, or a1, it'll usually be INPUT, as those are the three input bitSnap connectors on the module.

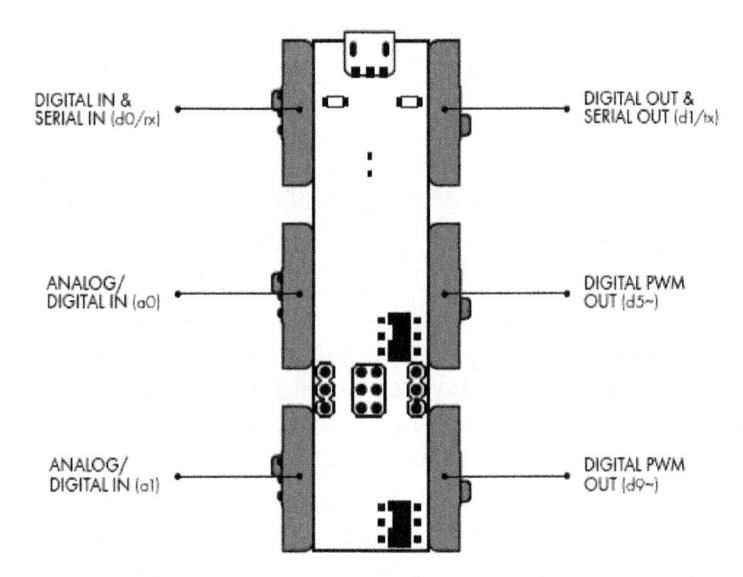

DIGITAL IN &
SERIAL IN (d0/rx)

ANALOG/
DIGITAL IN (a0)

ANALOG/
DIGITAL IN (a1)

DIGITAL OUT &
SERIAL OUT (d1/tx)

DIGITAL PWM
OUT (d5~)

DIGITAL PWM
OUT (d9~)

Figure 5-5. *The various inputs and outputs on the Arduino module.*

digitalWrite

`digitalWrite` is a function that sets an output to be ON or OFF. In the Arduino language, HIGH means ON and LOW means OFF. Be sure that the pin you're turning ON or OFF has been set as an OUTPUT with the function pinMode. You'll usually use this with pins d1, d5, or d9, the output bitSnap connectors. The syntax for `digitalWrite` is:

```
digitalWrite(pin, HIGH or LOW);
```

The first parameter in the parentheses is the pin that you're changing. The second sets that pin HIGH (ON) or LOW (OFF). The pin will hold that mode until it's changed with another `digitalWrite` statement.

delay

`delay` is a function that tells the microprocessor to stop and wait for some amount of milliseconds before moving onto the next statement. It's useful for when you want to hold the state of an output for some amount of time. For example, you can use it to turn an output ON and leave it that way for 1 second before turning it off. The syntax is:

```
delay(milliseconds);
```

The only parameter is the amount of time to wait. For example to stop and wait for one second before executing the next line of code, you'd use the statement:

```
delay(1000);
```

`delay` is function that you'll encounter a lot, but keep in mind that when you execute `delay`, nothing else will happen until the delay is over. That means you'll have to do some clever coding if you want your outputs to blink on different time periods.

Now that you have a deeper understanding of `pinMode`, `digitalWrite`, and `delay`, try modifying Example 5-1 so that all three outputs blink, and upload it to your board.

Arduino Inputs and Outputs

As you learned the basics about sketches, you started to cover some of the groundwork for outputs with Arduino. There are a few more things to know about outputs and there is the whole realm of inputs to discover. To learn about these topics, first upload a new code example (Example 5-3) to the board. When you push the button, the bargraph should start to fill up and then drop back down to zero. Or if you have an LED Bit, it will fade up and then turn off immediately.

1. Connect a button to d0 and an LED or bargraph Bit to d5. Also connect a power Bit. Your setup should look something like Figure 5-6.

2. Next to the d5 bitSnap connector there's a switch, make sure it's set to "analog."

3. Click File→New to create a new sketch.

4. Enter the code from Example 5-3 into the code window.

5. Go to Tools→Board and make sure "Arduino Leonardo" is still selected.

6. Go to Tools→Serial Port and make sure you've got the right serial port selected.

7. Click the Upload button on the toolbar.

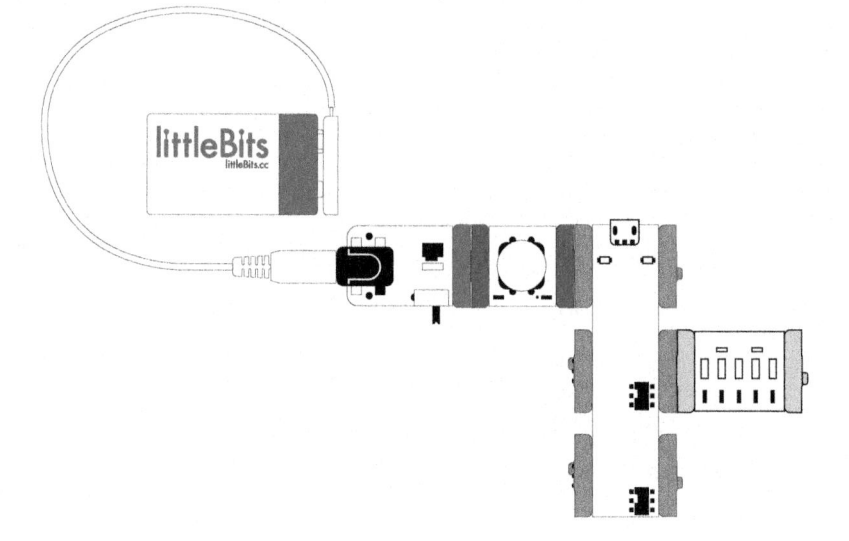

Figure 5-6. *Using inputs and analog outputs.*

In your code, when you're referring to the input pins, you'll simply use "0" for the d0/rx connector, A0 for the connector labeled a0 and A1 for the connector labeled a1. As for the output pins d1/tx, d5, and d9, you'll refer to them as 1, 5, and 9 respectively in your code.

Example 5-3. Inputs and Outputs

```
void setup() {
  pinMode(0, INPUT);  ❶
  pinMode(5, OUTPUT);
}

void loop() {
  if ( digitalRead(0) == HIGH ) {  ❷
    for (int i = 0; i <= 255; i++) {  ❸
      analogWrite(5, i);  ❹
      delay(10);
    }
    delay(500);
    analogWrite(5, 0);  ❺
  }
}
```

❶ Set up pin d0 as an input and d5 as an output.

❷ Execute the code in the curly braces if d0 is ON.

❸ Use a for loop to execute the block of code within the curly braces 256 times, counting 0 to 255 and storing that value in a variable called i each step of the way.

❹ Set the analog output value of pin d5 to the current value of i.

❺ Set the analog output value of pin d5 to 0.

There are a lot of new things to learn in Example 5-3. Let's break it down.

digitalRead

Example 5-3 introduces you to `digitalRead`, which checks for the state of a pin and then lets your code take action appropriately. The syntax of `digitalRead` is:

```
digitalRead(pin);
```

The `digitalRead` function returns the value HIGH or LOW, depending on the state of that input. When the button is pressed, it will return HIGH. When it's not pressed it will return LOW. When used in conjunction with an *if statement*, you can set up a block of code to execute only if it senses the pin is HIGH:

```
if ( digitalRead(0) == HIGH ) {
  // Any code between the curly braces will be executed
  // when the pin is HIGH.
}
```

analogWrite

`analogWrite` also makes an appearance in Example 5-3. As opposed to `digitalWrite`, which can only turn a pin ON or OFF, `analogWrite` can set a pin to a range of values from totally off to totally on. On the Arduino module, only the output pins d5 and d9 are capable of outputting an analog signal. These pins are marked with a tilde (~) to remind you that they're capable of analog output. Here's the syntax for `analogWrite`:

```
analogWrite(pin, value from 0 to 255);
```

Of course, the first parameter is the pin that you want to control and the second parameter takes a value from 0 (totally off) to 255 (totally on).

Both bitSnap connectors that are capable of analog output have a switch to change them between PWM and analog (Figure 5-7). Since the chip on the Arduino is digital, it's only capable of outputting signals that are either totally off (0 volts) or totally on (5 volts). However, it has the capability of simulating an analog output by pulsing the pins really quickly. This is called *pulse width modulation*. When you switch the output from PWM to analog, it runs the PWM signal through a *low pass filter*, which converts those pulses into a steady DC voltage in the range of 0 volts to 5 volts.

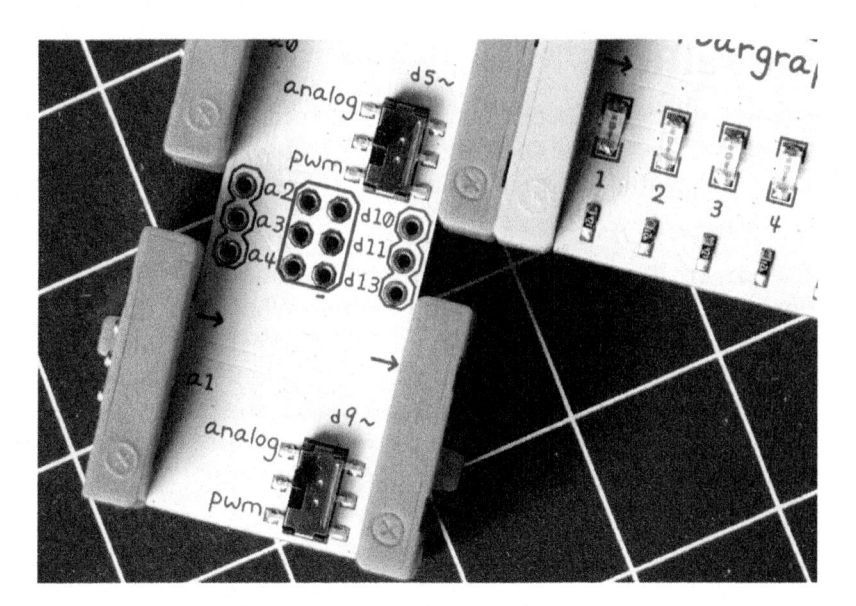

Figure 5-7. *The switches on d5 and d9, which are capable of analog output control whether the pins are pulsed (PWM) or if there's variable voltage (analog).*

You can see the difference between the two modes with the bargraph module. In PWM mode, it's pulsing the pins between 0 volts and 5 volts, so it gives the effect of all the LEDs on the module fading up in unison. Switch it to analog mode, and you'll get the expected behavior from the bargraph, with each LED lighting up in succession as the voltage goes from 0 to 5 volts.

For most projects, it's best if you keep the switches in analog mode. But if you'd like to use the Arduino module as a signal generator in a Synth Kit project, put the switches in PWM mode. See OK GO "Another Set of Objects" Synth (*http://littlebits.cc/projects/ok-go-another-set-of-objects-synth*) by Paul from littleBits in collaboration with OK GO for an example of using the Arduino module with Synth Bits.

In Example 5-3, the value of the analog output pin is gradually changed, controlled by a for loop. A for loop is frequently used to repeat a block of code a certain number times (though there's also so much more it can do). The syntax is a little tricky:

```
for (int i = 0; i <= 255; i++) {
    // Any code between the curly braces will be executed
    // 256 times
}
```

Within the preceding block of code, you can use the variable i to get the value of the current iteration. That is to say, the first time the loop runs, i will equal 0, the second time i will equal 1, and so on until i equals 255. This is why i is used in the analogWrite function call in Example 5-3. Every time the loop runs, the analog value increases until it's fully on. After the loop has iterated 256 times, the next line of code after the loop is run. In the case of Example 5-3, it waits for a half of a second and then sets the analog output of pin 5 back to 0.

Now that you know digital output, digital input, and analog output, that leaves analog *input* to try out:

1. Connect a dimmer to a0.
2. Connect an LED or another light up output Bit to d1.
3. Connect a power Bit. Your setup should look something like Figure 5-8.
4. Create a new sketch with the code from Example 5-4 and upload it to the Arduino Bit. This sketch will flash the LED at a rate determined by the setting of the analog input.

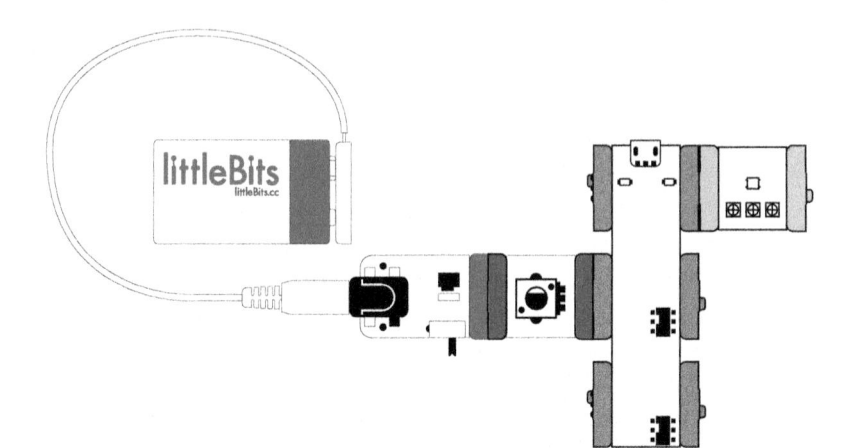

Figure 5-8. *Adding an analog input Bit to pin a0.*

Example 5-4. Analog Input

```
void setup() {
  pinMode(1, OUTPUT);
}

void loop() {
  int n = analogRead(0);  ❶
  digitalWrite(1, HIGH);
  delay(n);  ❷
  digitalWrite(1, LOW);
  delay(n);  ❸
}
```

❶ Create a variable called n and store the value from the analog input pin 0 in it.

❷ After turning pin d1 ON, wait n milliseconds.

❸ After turning pin d1 OFF, wait n milliseconds.

analogRead

Example 5-4 introduces how to get the analog value from an input. On the Arduino module, there are two bitSnap connectors that are capable of reading analog input: a0 and a1.

When you're using a0 and a1 as analog inputs, it's not necessary to use the `pinMode` function to set them as inputs.

Here's the syntax to use when you call `analogRead`:

```
analogRead(pin);
```

This function will return a value from 0 to 1023, with 0 being the equivalent of 0 volts (entirely off) and 1023 being the equivalent of 5 volts (entirely on). Example 5-4 stores the value of the `analogRead` statement into a new integer variable called n. A *variable* is simply a spot in the chip's memory to store data. An integer variable stores whole numbers.

To get the value stored in the variable, you just use the name of the variable wherever it's needed. So instead of calling `delay` with a specific number of milliseconds, place the name of the variable between the parentheses and the value of that spot in memory will be the amount of milliseconds it waits.

Keyboard and Mouse Control

One of the fantastic features on the chip that powers the Arduino module is that it can act like a USB keyboard or mouse. This means that your projects can easily interact with your computer. If you want to make a custom computer game controller out of littleBits, you can do just that.

To try making a simple mouse controller with your Bits:

1. Connect a button to d0 on the Arduino module.
2. Connect one slide dimmer to a0 on the Arduino module and a second slide dimmer to a1.
3. Connect the button and two slide dimmers to a power module using wires and a fork or branch. (See Figure 5-9.)
4. Set the slide dimmers to their center position.
5. Connect your Arduino module to your computer and upload the code in Example 5-5.

Now when you connect your project to your computer and power it on, your computer will think it's a mouse. Sliding the dimmers will move the pointer around and the button will trigger a single mouse click.

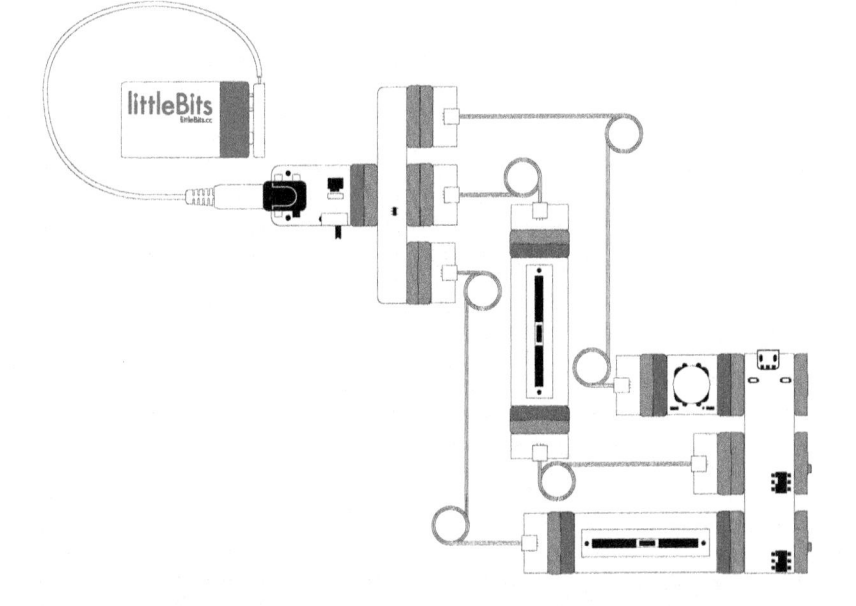

Figure 5-9. *Using a button and two sliders to make a custom mouse controller.*

Example 5-5. Mouse Controller

```
int xPos;   ❶
int yPos;   ❷

void setup() {
  Mouse.begin();   ❸
  xPos = analogRead(A1);   ❹
  yPos = analogRead(A0);   ❺
  pinMode(0, INPUT);
}

void loop() {
  int xMove = analogRead(A1) - xPos;   ❻
  int yMove = analogRead(A0) - yPos;   ❼
  Mouse.move(xMove, yMove);   ❽
```

```
    xPos = xPos + xMove;   ❾
    yPos = yPos + yMove;   ❿
    if (digitalRead(0) == HIGH) {
      Mouse.click(MOUSE_LEFT);   ⓫
    }
}
```

❶   Create a global variable called **xPos**, which will store the x position (left/right) of the mouse.

❷   Create a global variable called **yPos**, which will store the y position (up/down) of the mouse.

❸   Begin mouse emulation.

❹   Read the initial value of analog input a1 and store it in the global variable **xPos**.

❺   Read the initial value of analog input a0 and store it in the global variable **yPos**.

❻   Create a local variable, **xMove**, and store the difference between the current value of analog input a1 and the last reading.

❼   Create a local variable, **yMove**, and store the difference between the current value of analog input a0 and the last reading.

❽   Move the mouse left or right some number of pixels based on the value of **xMove**.

❾   Update **xPos** to reflect the new position.

❿   Update **yPos** to reflect the new position.

⓫   If the button is pressed, emulate a left mouse click.

This example introduces a few new concepts, *variable scope* and mouse functions.

Variable Scope

Example 5-5 highlights the concept of variable scope in Arduino programming. The two variables created at the top, xPos and yPos, are *global variables*: they can be read and written from anywhere in the sketch since they're defined outside of any code blocks, which are enclosed by curly braces.

On the other hand, xMove and yMove are defined within the loop function and can only be read or written within that function. After the loop completes, those variables are destroyed (to be created again the next time the loop executes). xMove and yMove are *local variables*.

Mouse Functions

This example also uses the mouse functions.

Mouse.begin

This function is used to tell the Arduino module to start emulating the mouse. Typically, you'll call this within your setup function without any parameters:

```
Mouse.begin();
```

Mouse.move

This function tells your Arduino to emulate the movement of the mouse some number of pixels on the x axis and the y axis. It's called with the x value first and the y value second

```
Mouse.move(x,y);
```

Negative x values move the mouse to the left. Positive x values move the mouse to the right. Negative y values move the mouse up. Positive y values move the mouse down.

Mouse.click();

This tells your Arduino to emulate a single click and release of the mouse. The only parameter is which button to click:

```
Mouse.click(MOUSE_LEFT);
```

Other options include: MOUSE_RIGHT and MOUSE_MIDDLE.

There are a few other mouse functions that might come in handy. See Arduino's Mouse and Keyboard reference (*http:// arduino.cc/en/Reference/MouseKeyboard*) for more information.

Project: HelloRun Game Controller

http://littlebits.cc/projects/hellorun-controller

There's a fun and free web-based game called HelloRun (*http://hellorun.helloenjoy.com/*). You fly an aircraft through a maze, navigating around obstacles with your keyboard with the up and down keys. In this project, you'll use Legos, roller switches, and the Arduino module to make a custom controller for the game.

The exact way you build the controller using Legos is up to you, but there are a few important parts you'll definitely need:

- 1 Arduino Bit
- 2 roller switch Bits
- 1 split Bit
- 2 wire Bits
- 2 littleBits brick strips, socket type
- 1 2x2 Lego plate with inverted snap (part 4237084)
- 1 1x2 Lego Technic brick (part 370001)
- 2 1x2/2x2 Lego Angle plate (part 4278046)

Here's how to make your own:

1. Connect a split module to the power module and connect
those to the inputs of the roller switches.

2. Connect the roller switches to the first two inputs of the
Arduino module. The switches should be in "close" mode.
Figure 5-10 shows the circuit.

3. Program the Arduino module with the code in Example 5-6
and check that the actuation of the roller switches sends an
up or down key press.

4. Build up a 2x8 Lego wall from the building plate.

5. Connect the Lego angle plate on top so that it holds the
Brick Adapters on the side of the wall. Attach the roller
switches (in *close* mode) to the angle plates.

6. Place the 2x2 Lego plate with inverted snap into the Technic
brick and place the Technic brick on top of the wall.

7. Place a 2x8 Lego brick onto the rotating 2x2 plate so that the
2x8 brick actuates the roller switches. This might take a little
bit of trial and error.

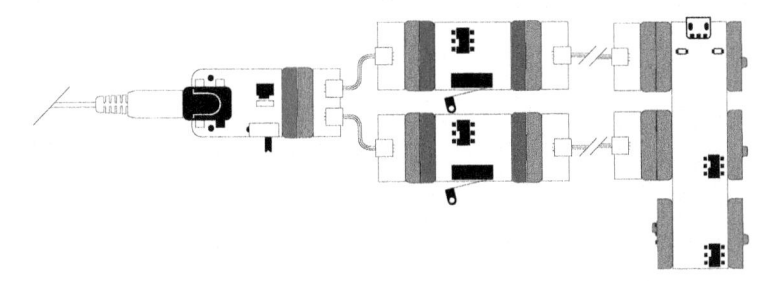

Figure 5-10. *HelloRun diagram*

Example 5-6. HelloRun Controller

```
void setup() {
  Keyboard.begin();  ❶
  pinMode(0, INPUT);
  pinMode(A0, INPUT);
}
void loop() {
  if (digitalRead(0) == HIGH) {
    Keyboard.press(KEY_UP_ARROW);  ❷
    while (digitalRead(0) == HIGH) {
      // Do nothing until the switch is released
    }
    Keyboard.releaseAll();  ❸
  }
  if (digitalRead(A0) == HIGH) {
    Keyboard.press(KEY_DOWN_ARROW);  ❹
    while (digitalRead(A0) == HIGH) {
      // Do nothing until the switch is released
    }
    Keyboard.releaseAll();  ❺
  }
}
```

❶ Begin keyboard emulation.

❷ Press and hold the up key.

❸ When the roller switch is released, release the key.

❹ Press and hold the down key.

❺ When the roller switch is released, release the key.

Figure 5-11. *The controller uses two inputs on the Arduino module.*

Figure 5-12. *The Brick strips help you attach the roller switches to the side of the wall.*

This project introduces two new concepts, keyboard emulation and the while loop.

Keyboard Emulation

Keyboard emulation works a lot like mouse emulation.

Keyboard.begin

This function is used to tell the Arduino module to start emulating the keyboard. Typically, you'll call this within your setup function without any parameters:

```
Keyboard.begin();
```

Keyboard.press

This function tells the Arduino to press and hold a particular key. Here's the syntax:

```
Keyboard.press(KEY);
```

For example, to type a lowercase letter n, you'd use:

```
Keyboard.press('n');
```

For a list of special keys, including the modifiers like shift and control, see the Arduino reference page (*http://arduino.cc/en/Reference/KeyboardModifiers*) on the topic.

Keyboard.releaseAll

This function tells the Arduino to let go of any pressed keys. It's used without any parameters:

```
Keyboard.releaseAll();
```

There are a few other keyboard functions that you'll find useful. See Arduino's Mouse and Keyboard reference (*http://ardu ino.cc/en/Reference/MouseKeyboard*) for more information.

while

The while keyword means that the block of code in curly braces is to be executed as long as some condition is true:

```
while (condition) {
        // Keep looping this code as long as condition is true.
}
```

In the case of Example 5-6, it's meant to hold the key down until the roller switch is released.

Using the Arduino Bit with Scratch

Scratch (*http://scratch.mit.edu*) is a visual tool for creating programs, stories, games, and animations (Figure 5-13) . It comes from the Lifelong Kindergarten Group at the MIT Media Lab and is designed to help children "think creatively, reason systematically and work collaboratively." You can use Scratch to easily program Arduino Bits for projects such as the Wind Turbine Simulator (*https://www.youtube.com/watch?v=_-apSTKW3jE*), which simulates collecting and using wind power. Check out the littleBits Scratch extensions (*http://littlebits.cc/education-scratch-extensions*) to get started using Scratch with your Bits.

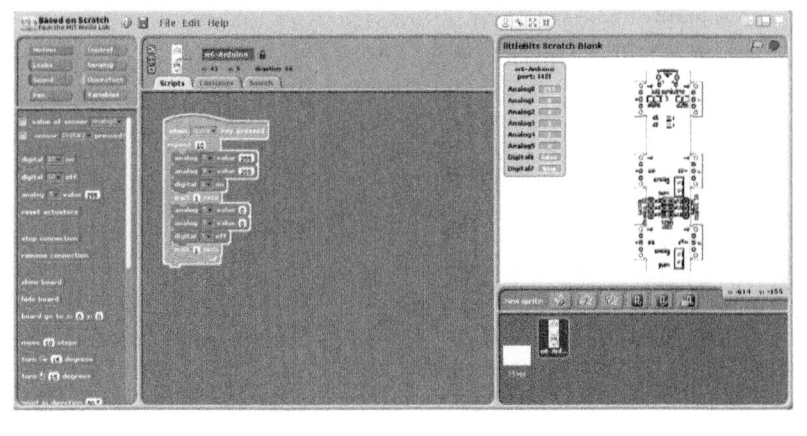

Figure 5-13. *Scratch and littleBits together*

6/Making Your Own Bits

Although there's a huge library of Bits available for you to choose from, littleBits wants to make the library infinite! There are a few tools that help you build circuits to create your own Bits. This chapter covers some of these prototyping tools and tells you how you can even sell your Bits online on the Bitlab (*http://littlebits.cc/ bitlab*)—the app store for hardware.

Before we introduce the tools that help you create your own Bits, it would be helpful to review some of the technical specifications of the littleBits platform. Let's start by taking another close look at the bitSnap connector, shown in Figure 6-1.

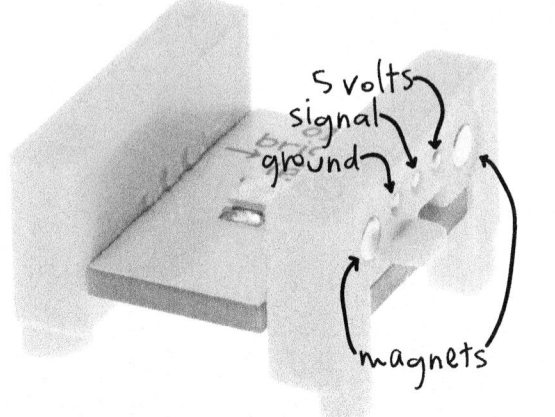

Figure 6-1. *The connectors on the bitSnap*

Between the two magnetic pads, there are three electrical connections that go between each Bit. Every Bit has a connection to 5 volts and to ground through the two outside terminals on the BitSnap connector. This is how every Bit is powered.

The middle terminal carries the *signal*, which can be anywhere between 0 and 5 volts. For digital signals, 0 volts represents OFF, whereas 5 volts represents ON. For analog signals, the signal can be 0 volts, 5 volts, or anywhere in between.

As you prototype, it is helpful to have a tool called a *multimeter*, which can measure the voltage, among other possible characteristics in a circuit. You don't need a fancy multimeter, as they can get quite expensive. A basic, $20 multimeter from Maker Shed, Adafruit, or Sparkfun will serve you well for a long time.

If you don't have a multimeter, you can always use the number Bit in voltage mode or a bargraph Bit to sense the voltage of the signal in your littleBits circuit as you prototype.

Proto Module

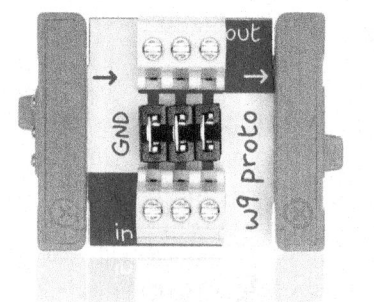

The proto module will look quite different than any of the other Bits you've used previously. It has two sets of three screw terminals and a set of three jumpers. The screw terminals are what allow you to connect your own wiring to a littleBits circuit. The jumpers bridge the connections between the input bitSnap connector and the output bitSnap connector.

With all of its jumpers in place, it works like a basic Wire module, passing the electricity and the signal along to the next Bit.

You can prove that quite easily:

1. Attach the proto module with its jumpers in place to a power Bit.

2. Attach any output Bit to the proto module.

   As you'll see, it passes along the electricity and the ON signal from the power module, actuating the output Bit just like a wire Bit would.

3. Pull out the jumper on the left side, next to the label GND (ground).

When you pull the left-most jumper out, the connection to ground is broken and the circuit is incomplete.

4. Replace the ground jumper.

5. Pull out the jumper on the right side.

   When you pull the right-most jumper out, the connection to 5 volts is broken and again, the circuit is incomplete.

6. Replace the 5 volt jumper.

In most cases, you'll want to keep the ground and 5 volt jumpers in place. If you're creating an output Bit, you'll also keep the signal jumper in place because output Bits pass along the signal to the following Bits. If you're creating an input Bit, you'll remove the signal jumper and your circuit will have some effect on the signal before it's passed along to the next Bit.

The proto module is included in the Hardware Development Kit (*http://littlebits.cc/kits/hdk*), along with a perf module (w29) and bitSnap connectors. The perf module allows you to build a small circuit right on the module. When you're ready to design your own PCB, you can use the gray bitSnaps (Figure 6-2) to attach to the board. The included color-coded stickers let you easily denote your bitSnaps as power, input, wire, or output.

Figure 6-2. *Gray bitsnaps*

Let's start with making a simple output Bit.

Creating an Output

Outputs show what's happening on the signal line in addition to passing along the signal to the next Bit. In this section, you'll make a very simple output Bit with an LED.

To try out the example in this section, you'll need a few electronics prototyping tools and supplies:

- breadboard
- hookup wire or jumper wire
- wire strippers
- assorted LEDs
- assorted resistors

To make your own output:

1. Connect a dimmer module to a power module.
2. Connect the Proto module to the dimmer. All three jumpers should be in place.
3. Insert a jumper wire into the center position of the IN screw terminal and screw it down. This is the input signal from any bits that precede the Proto module *to* your output.
4. Insert a jumper wire into the left position of the IN screw terminal (GND, or ground) and screw it down.
5. Connect both pieces of hookup wire to connect to separate vertical rows in the breadboard.
6. Place an LED into the bread board so that the shorter lead (the *cathode*) is in the same vertical row as the wire from ground.
7. Place a resistor into the breadboard so that it connects the longer lead of the LED (the *anode*) to the wire from the signal line. In most cases, a 150 ohm resistor will work, but see "Choosing the Right Resistor" on page 152 to learn how this value is derived. Your setup should look something like Figure 6-3.

Adjust the dimmer, and watch the LED change.

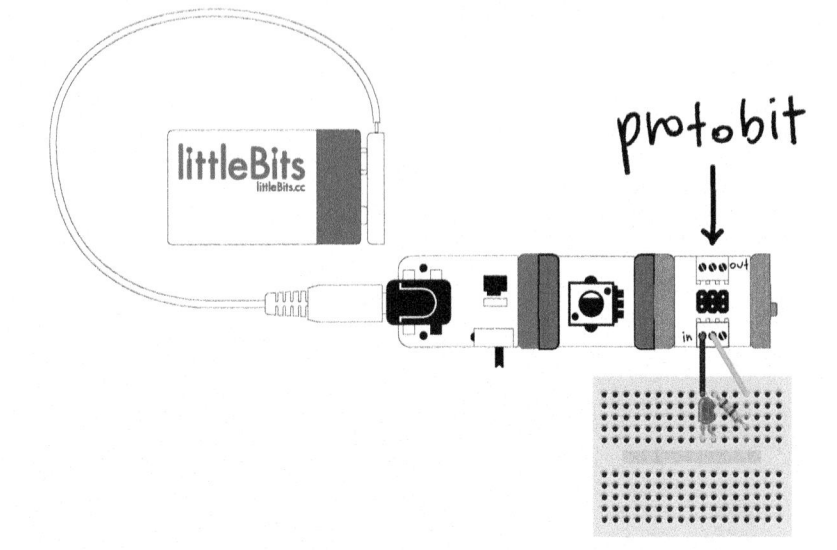

Figure 6-3. *Making an output with the Proto module. Bread-board diagram created in Fritzing (http://fritzing.org)*

Choosing the Right Resistor

It's important that you use a *current limiting resistor* in your LED circuit so that you don't pull more current than needed and you don't damage your LED.

In most cases, a 150 ohm resistor will make a fine current limiting resistorfor typical LEDs, but here's how you can check.

There's some easy math to figure out the value of the resistor to use, which depends on the voltage and a couple of characteristics of the LED.

R = (Vs - VL) / IL

R
> Resistance in ohms

Vs
> Source voltage. For littleBits, it will always be 5 volts.

When your Bit Module is done, you can can submit it to bitLab and if enough of the community votes for your project, you can get your Bit Module made. Here are some examples of successful bitLab Bits

ScopeBit
   http://littlebits.cc/bitlab/bits/oscilloscope

   The Oscilloscope Module (or ScopeBit) is a small output module that displays the waveform of the signal going through it.

EMG SpikerBox
   http://littlebits.cc/bitlab/bits/emg-spikerbox

   The EMG SpikerBox module detects the electrical activity of human muscles non-invasively using simple skin surface electrodes.

MaKey MaKey Module
   http://littlebits.cc/bitlab/bits/the-makey-makey-module

   The MaKey MaKey turns everyday objects into touchpads and combines them with computer programs and the internet.

Creating an Input

An input modifies the signal line that's passing through it to the next Bit. In this section, you'll make a very simple input Bit with a pushbutton switch.

To try out the example in this section, you'll need a few electronics prototyping tools and supplies:

- breadboard
- hookup wire or jumper wire
- wire strippers
- momentary tactile switch, such as Adafruit part 367 (*http://www.adafruit.com/product/367*) or the switches included with Maker Shed's Mintronics: Survival Pack (*http://www.makershed.com/products/mintronics-survival-pack*).

To make your own output:

1. Connect the Proto module to a power module.
2. Leaving the left and right jumpers in place, remove the center jumper from the Proto module. This will prevent the power Bit's signal from passing through the Proto module automatically.
3. Insert a jumper wire into the center position of the OUT screw terminal and screw it down. This is the input signal *from* your input to any bits that follow the Proto module.
4. Insert a jumper wire into the center position of the IN screw terminal (signal from any bits that precede yours) and screw it down.
5. Place a button into the breadboard so that it straddles the center of the breadboard. If it doesn't fit correctly, rotate the button 90 degrees.
6. Connect one wire from the Proto module to the leftmost vertical row that the button is connected to, and the other wire to the rightmost one.
7. Connect an output, such as an LED Bit, to your Proto module. Your setup should look something like Figure 6-4.

Press the button and watch the led Bit light up.

Signal Override

Try putting a button Bit between the power Bit and your Proto module. You need to press both buttons to get the signal to pass to the output Bit.

What if you want a button that doesn't care about the state of the signal coming from the bit immediately before it? To do this, disconnect the wire from the center position of the IN screw terminal and connect that wire to the right position of the OUT screw terminal (power). That way, whenever you press your button, it will transmit a signal no matter what signal it gets from the Bits before it.

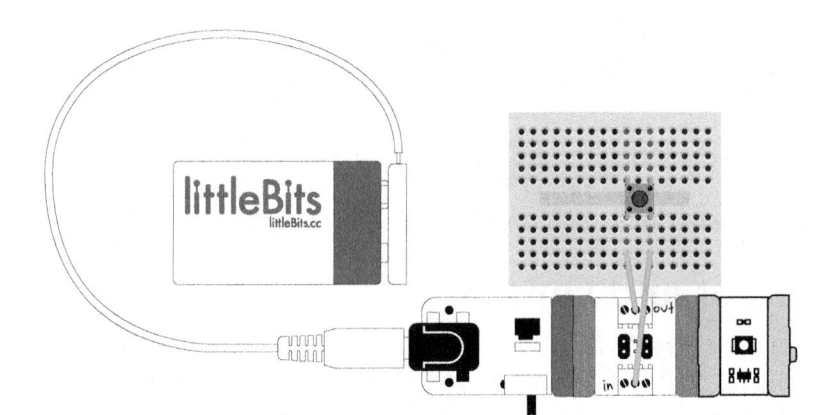

Figure 6-4. *Making an input with the Proto module. Breadboard diagram created in Fritzing (http://fritzing.org)*

Perf Module

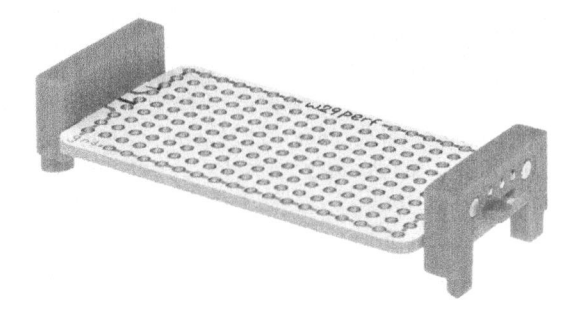

The Perf module is perfect for building temporary circuits. If you've created a circuit that you want to make permanent, you can solder that circuit onto the Perf module.

Input and Output

Let's use the Perf module to make a Bit that's both an input and an output. It will light up an LED when it receives a signal, and it will have a button that you can press to send a signal. We'll wire it up as described in "Signal Override" on page 155, so you can send a signal even if a preceding Bit hasn't sent this Bit a signal, which is different from the standard button Bit.

To try out this example, you'll need all the parts and tools you used in this chapter's earlier projects.

1. Look closely at Figure 6-5 and arrange the components on the Perf module as shown. Here are few things to pay attention to:

   a. You're connecting the resistor from the signal to a through-hole that's not connected (yet) to anything.

b. You're connecting the short lead of the LED to the bottom row of the Perf module, which is labeled "gnd" on the module. The long lead goes to a through-hole that's not connected to anything else.

c. You need to connect the momentary tactile switch in such a way that pushing the button will send 5 volts to the output signal. To do this, you'll need to rotate it 90 degrees from the way you had it in the breadboard in the previous example. This is because, as oriented in Figure 6-7, the two bottom leads of the button are connected internally to each other, and the two top leads are also connected to each other. It's only when you press the button that the top and bottom leads are connected, making the connection between power and the signal line.

2. Flip the Perf bit over, and bend the long lead of the LED so it makes contact with the resistor's leads. Make the connections shown in Figure 6-6 with a soldering iron and solder that's designed for use with electronics.

3. Check your connections closely. Connect a power Bit, button Bit, Perf module, and LED Bit as shown in Figure 6-7.

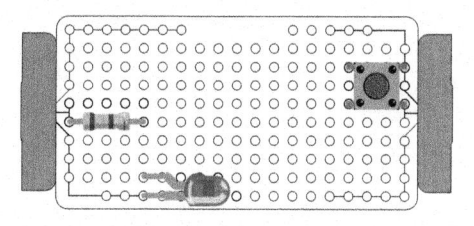

Figure 6-5. *Making an combined input and output with the Perf module. Button, LED, and resistor diagrams from Fritzing (http:// fritzing.org)*

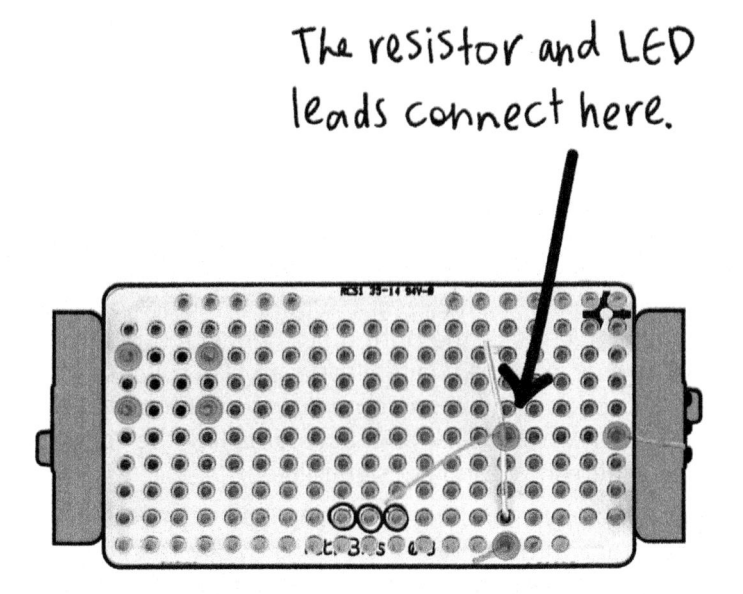

The resistor and LED leads connect here.

Figure 6-6. *Here's where to make the solder connections.*

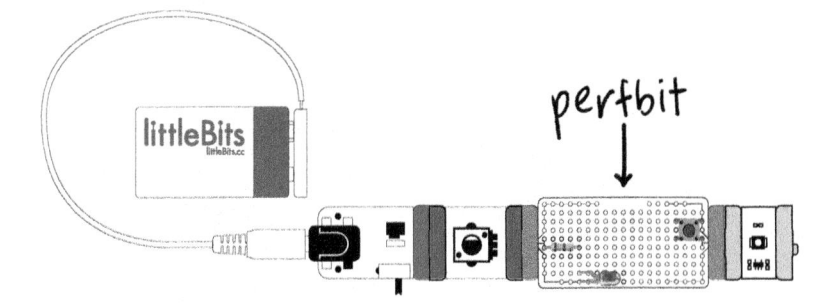

perfbit

Figure 6-7. *Connecting the Perf module*

With everything hooked up, turn on the power Bit. If all went well, the LED should light up when you press the button Bit. The LED Bit will light up when you press the button on the Perf module.

What Can You Make?

Between the Proto module and the Perf module, you can make all kinds of inputs and outputs for your littleBits projects. Here are a few that show the possibilities:

Humidity Sensor
> bltRobotics created this project for the bitLab (*http://little bits.cc/bitlab*), where littleBits customers can vote on users' hardware creations. The winning projects could end up getting made by littleBits. The Humidity Sensor measures relative humidity and outputs an analog signal between 0 and 5 volts.

Touch Sensor
> The touch sensor was created for the bitLab by Bare Conductive. It is a powerful capacitive proximity sensor that sends a signal to the attached Bits as you physically approach it. The closer you get, the higher the voltage it sends out. Use an alligator lead with one end attached to the gold electrode on the touch sensor, then connect the other end to something conductive like tin foil, conductive foam, or even paint a square of Electric Paint. Once turned on, the touch sensor will automatically calibrate the material to create a custom capacitive sensor off of the module. Fine tune your proximity range using the included screwdriver and use it to detect a cookie thief with the Buzzer or have hands-free control of the Oscillator from the Synth Kit to drive some sweet audio!

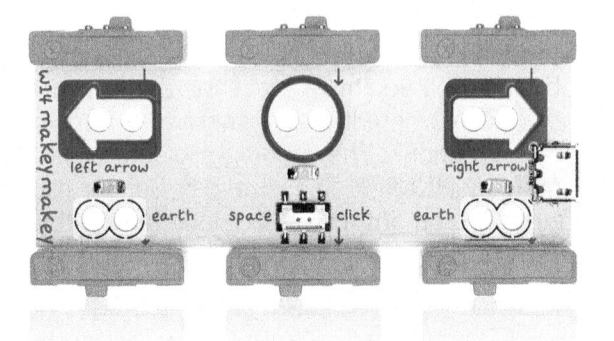

The MaKey MaKey turns everyday objects into touchpads and combines them with computer programs and the internet. Now, you can trigger your littleBits modules with everyday objects or use littleBits modules to trigger events in your computer, or a combination of both.

The module connects to a computer through a micro USB cable and has three MaKey MaKey inputs which are mapped to left arrow, right arrow, and space bar/mouse click (depending on how you set the switch). Each one of these key inputs can be controlled by littleBits modules like motion triggers or light sensors. The MaKey MaKey inputs are also connected to littleBits outputs so you can use a banana to control an oscillator or a glass of water to move a servo motor.

The Bleep Drum

The Bleep Drum is a lo-fi Arduino-based drum machine. The littleBits module version can be triggered from any changing signal, like the output from an oscillator, sequencer, or even a simple button. These crunchy samples go great with the smooth sounds of the synth modules!

The Bleep Drum module has two modes. In "analog" mode, the drum sample and its pitch are determined by an incoming analog level like that of the sequencer module. In "digital" mode the pitch and sample are selected by the Bleep Drum's knob. The sample is activate every time a pulse is received from any module, from a button, light sensor, oscillator, or anything in between.

There's much more you can make with these modules. Visit the littleBits Hardware Development kit (*http://littlebits.cc/kits/hdk*) page and the bitLab forum (*http://discuss.littlebits.cc/c/bitlab*) for the latest on what you can do—and what others are making—by combining creativity, electronics, and littleBits!

Index

W

wall wart, 4
waveform, 65
while loop, 145
wire
    Bits, 17
wire Bit, 18
wireless communication, 95-117

wireless receiver Bit, 95
wireless transmitter Bit, 95

X

XOR Bit, 59

Z

Zoetrope project, 36

About the Authors

Ayah Bdeir is the founder and CEO of littleBits, an award-winning library of electronic modules that snap together with magnets to allow anyone to learn, build, and invent with electronics. Bdeir is an engineer, interactive artist and one of the leaders of the open source hardware movement. Bdeir's career and education have centered on advancing open source hardware to make education and innovation more accessible to people around the world. She is a co-founder of the Open Hardware Summit, a TED Senior Fellow and an alumna of the MIT Media Lab. Bdeir was named one of Inc. Magazine's 35 Under 35, one of NY Business Journal's Women of Influence, one of Fast Company's 100 Most Creative People in Business, one of Popular Mechanics' 25 Makers Who Are Reinventing the American Dream, one of Entrepreneur's 10 Leaders to Watch, one of the CNBC Next List, and one of MIT Technology Review's 35 Innovators Under 35. Originally from Lebanon and Canada, Ayah now lives in New York City.

Matt Richardson is a San Francisco-based maker and author. He's the owner of Awesome Button Studios, a consultancy focused on blending creativity and technology. After graduating with a Master's from New York University's Interactive Telecommunications Program (ITP) in 2013, he continued his work there as a resident research fellow. Matt is the co-author of *Getting Started with Raspberry Pi* and the author of *Getting Started with BeagleBone* and *Getting Started with Intel Galileo*.

Colophon

The cover photo is by littleBits Electronics, Inc. The cover fonts are Benton Sans and Soho. The text font is Benton Sans; the heading font is Serifa; and the code font is The Sans Mono.